AF725885

Steel Water-Storage Tanks

AWWA MANUAL M42

Revised Edition

American Water Works
Association

Manual of Water Supply Practices — M42, Revised Edition

Steel Water-Storage Tanks

Copyright © 2013 American Water Works Association

All rights reserved. No part of this publication may be reproduced or transmitted in any form or by any means, electronic or mechanical, including photocopy, recording, or any information or retrieval system, except in the form of brief excerpts or quotations for review purposes, without the written permission of the publisher.

Disclaimer
The authors, contributors, editors, and publisher do not assume responsibility for the validity of the content or any consequences of its use. In no event will AWWA be liable for direct, indirect, special, incidental, or consequential damages arising out of the use of information presented in this book. In particular, AWWA will not be responsible for any costs, including, but not limited to, those incurred as a result of lost revenue. In no event shall AWWA's liability exceed the amount paid for the purchase of this book.

AWWA Sr. Manager of Editorial Development and Production: Gay Porter De Nileon
AWWA Sr. Technical Editor/Project Manager: Martha Ripley Gray
Cover Art: Melanie Yamamoto, AWWA Sr. Graphic Designer
Production: Sandra Lankenau, AWWA Sr. Production Editor

Library of Congress Cataloging-in-Publication Data

Steel water-storage tanks.—Revised edition.
pages cm.—(AWWA manual ; M42)
Includes bibliographical references and index.
ISBN 978-1-58321-948-5 (alk. paper)
1. Water towers—Design and construction—Handbooks, manuals, etc. 2. Water towers--Maintenance and repair--Handbooks, manuals, etc. 3. Tanks—-Design and construction—-Handbooks, manuals, etc. 4. Tanks—Maintenance and repair—Handbooks, manuals, etc. I. American Water Works Association.
TD489.S74 2013
628.1'3--dc23

2013012693

Printed in the United States of America

ISBN-13 978-1-58321-948-5 eISBN-13 978-1-61300-238-4
ISBN-10 1-58321-948-X eISBN-10 1-61300-238-6

American Water Works Association

6666 West Quincy Avenue
Denver, CO 80235-3098
303.794.7711
www.awwa.org

Printed on recycled paper

Contents

This page intentionally blank.

Figures

This page intentionally blank.

Tables

This page intentionally blank.

Preface

The purpose of this manual is to provide the water distribution system manager, operator, contractor, and consultant with information concerning steel tanks used for the storage of water during water production, treatment, and distribution. The manual covers the planning, specification, construction, operation, and maintenance of steel tanks of riveted, welded, and bolted construction.

This manual is not intended to be a technical commentary on American Water Works Association (AWWA) standards dealing with steel water-storage tanks; reference is made to those standards to make the reader aware of their existence and application. ANSI/AWWA D101, Standard for Inspecting and Repairing Steel Water Tanks, Standpipes, Reservoirs, and Elevated Tanks for Water Storage, was last reaffirmed in 1986 and withdrawn as a standard in 1998, the year this manual was first published. That standard is included in this manual as appendix C.

While the scope of this edition of the manual remains largely unchanged, updates have been made to the content throughout. This revised edition includes several new and many revised figures and tables, information on composite elevated tanks, recommended steel tank guidelines, water quality assurance, references to AWWA standards, and current steel water tank industry sources and organizations.

This page intentionally blank.

Acknowledgments

The Standards Committee on Steel Elevated Tanks, Standpipes, and Reservoirs, which reviewed and approved this manual, had the following personnel at the time of approval:

Stephen W. Meier, Chair

User Members

J. Camarena, East Bay Municipal Utility District, Oakland, Calif. (AWWA)
T.M. Dawson, Emerald Coast Utilities Authority, Pensacola, Fla. (AWWA)
M. Gaffey, Naval Facilities Engineering Service Center, Washington, D.C. (AWWA)
*M.W. Griffin,** Missouri-American Water Co., St. Louis, Mo. (AWWA)
C.P. Harder, City of Fort Worth, Fort Worth, Texas (AWWA)
G.S. Jacobson, California Water Service Company, San Jose, Calif. (AWWA)
E.J. King, Connecticut Water Co., Clinton, Conn. (AWWA)
M.J. Mazurek, Baltimore County Department of Public Works, Towson, Md. (AWWA)
J.L. Ortiz, San Francisco Public Utilities Commission, San Francisco, Calif. (AWWA)
J.P. Scott, New Jersey–American Water Company, Voorhees, N.J. (AWWA)
J. Travers, City of Richardson, Richardson, Texas (AWWA)

General Interest Members

D.M. Algranti, Albert A. Webb Associates, Riverside, Calif. (AWWA)
J.D. Brock, Poggemeyer Design Group, Bowling Green, Ohio (AWWA)
E. Darrimon, Bay Area Coating Consultant Service Inc., Denair, Calif. (AWWA)
W.J. Dixon, Dixon Engineering, Inc., Lake Odessa, Mich. (AWWA)
M.L. Hickey, Murray, Smith & Associates Inc., Portland, Ore. (AWWA)
F.S. Kurtz,† Standards Engineer Liaison, AWWA, Denver, Colo. (AWWA)
S.W. Meier, Tank Industry Consultants, Indianapolis, Ind. (AWWA)
B.M. Shepherd,† Standards Council Liaison, Ocean Park, Wash. (AWWA)
J.I. Strand, Short Elliott Hendrickson Inc., Chippewa Falls, Wis. (AWWA)
C.C. Sundberg, CH2M HILL, Issaquah, Wash. (AWWA)
R.S. Wozniak, Bow Tech Ltd., Batavia, Ill. (AWWA)

Producer Members

J.W. Davis, Tnemec Company, Kansas City, Mo. (AWWA)
B.E. Kromer, Tank Builders Inc., Euless, Texas (AWWA)
G.A. Larson, CB&I Inc., Clive, Iowa (AWWA)
K. McGuire, CST Storage, Parsons, Kan. (AWWA)
R.L. Moore, Utility Service Company, Kirkwood, Mo. (AWWA)
L.D. Scott, Associated Construction & Engineering, Paso Robles, Calif. (AWWA)
*J. Snodgrass,** Utility Service Company, Omaha, Neb. (AWWA)
D.L. Stilger, Caldwell Tanks Inc., Louisville, Ky. (AWWA)

* Alternate
† Liaison

The AWWA Standards Committee on Steel Elevated Tanks, Standpipes, and Reservoirs would like to dedicate this manual to Robert E. Vansant, past chairperson of the Standards Committee on Steel Elevated Tanks, Standpipes, and Reservoirs, who initiated work on this manual.

The following organizations and companies generously provided photos and illustrative materials for use in this manual: A.O. Smith Harvestore Products Inc., Cathodic Protection Services Company, Chicago Bridge and Iron Company, GPE Controls, Harco Waterworks CP, Mid Atlantic Storage Systems Inc., North Consumer Products, PDM Hydrostorage, Peabody TecTank Inc., Resource Development Company, Steel Plate Fabricators Association, and Tank Industry Consultants Inc.

Introduction

More than 100,000 steel water-storage tanks have been constructed within the last 100 years, a value that far exceeds the number of large water-storage vessels of any other type of construction material. Many steel water tanks have service histories in excess of a century and are still in service today (Figure F-1). Whereas early tanks were riveted, modern practice uses welded or bolted design and construction, which provide the advantage of a zero leakage tolerance.

To further increase their potential service life, steel tanks can be dismantled and re-erected in new locations. A tank that was originally in an optimal location can become useless if a factory relocates or there is a shift in housing patterns. However, a steel tank can be dismantled and then erected and coated at a new location.

This manual provides information on the selection, design, construction, maintenance, inspection, and repair of steel tanks for potable water storage. The manual will assist in tank sizing, configuration, site selection, design, operation, and maintenance.

Source: Tank Industry Consultants Inc.

Figure F-1 A tank constructed in 1902 is still serving Wabash, Ind., with practically no metal loss

DEFINITIONS

The following definitions apply in this manual:

Bottom capacity level (BCL) The water level in the tank when the tank is emptied through the specified discharge fittings (unless otherwise specified by the purchaser). In an elevated tank, the elevation of the bottom capacity level(s) is as given by the purchaser and is determined by the design features of the tank configuration.

Capacity The net volume in gallons (liters) that may be removed from a tank emptied to its bottom capacity level after being filled to its top capacity level.

Constructor The party that furnishes the work and materials for placement and installation.

Elevated tank A container or storage tank supported on one or more columns.

Engineer An employee of the purchaser or, more commonly, a professional engineering firm engaged by the purchaser to perform specification and inspection services.

Head range The difference between the lower and upper capacity levels of a tank.

Manufacturer The person or company that furnishes the tank components.

Owner The person or firm that will own and operate the completed tank. The owner may designate agents, such as an engineer, purchaser, or inspector, for specific project responsibilities.

Purchaser The person, company, or organization that purchases the tank.

Reservoir A ground-supported, flat-bottom cylindrical tank with a shell height less than or equal to its diameter.

Standpipe A ground-supported, flat-bottom cylindrical tank with a shell height greater than its diameter.

Tank An elevated tank, standpipe, or reservoir used for water storage.

Top capacity level (TCL) The maximum operating level of water in a tank, as dictated by the elevation at which water discharges from the tank through the overflow pipe entrance. In a standpipe or reservoir, the top capacity level is as given by the purchaser. In an elevated tank, the elevation of the top and/or bottom capacity level(s) is as given by the purchaser.

AWWA STANDARDS

The majority of all steel potable water-storage tanks constructed in the United States adhere to specifications that reference the following American Water Works Association (AWWA) steel tank standards:

- ANSI/AWWA D100, Standard for Welded Carbon Steel Tanks for Water Storage.
- ANSI/AWWA D102, Standard for Coating Steel Water-Storage Tanks.
- ANSI/AWWA D103, Standard for Factory-Coated Bolted Carbon Steel Tanks for Water Storage.
- ANSI/AWWA D104, Standard for Automatically Controlled, Impressed-Current Cathodic Protection for the Interior Submerged Surfaces of Steel Water Storage Tanks.
- ANSI/AWWA D106, Standard for Sacrificial Anode Cathodic Protection Systems for the Interior Submerged Surfaces of Steel Water Storage Tanks.

- ANSI/AWWA D107, Standard for Composite Elevated Tanks for Water Storage.
- ANSI/AWWA D108, Standard for Aluminum Dome Roofs for Water Storage Facilities.
- ANSI/AWWA C652, Standard for Disinfection of Water-Storage Facilities.

These standards (except ANSI/AWWA C652) are developed and maintained under the direction of the AWWA Standards Committee on Steel Elevated Tanks, Standpipes, and Reservoirs, which is composed of members representing consumer (utility), general interest (academic and consulting engineering), and producer (constructor and manufacturer) groups. ANSI/AWWA C652 was developed and is maintained by the AWWA Standards Committee on Disinfection of Facilities.

Once a draft standard or revision is approved by a standards committee, it is forwarded to the AWWA Standards Council for review and approval. If approved by the council, it is offered for public review and then presented to the AWWA Board of Directors for final approval. The standards listed above have also been approved as standards by the American National Standards Institute (ANSI). AWWA D101-53 (R86), Standard for Inspecting and Repairing Steel Water Tanks, Standpipes, Reservoirs, and Elevated Tanks for Water Storage (included here as appendix C), was withdrawn upon publication of the first edition of this manual in 1998.

WELDED TANKS

Welded tanks have been used for water storage since the 1930s. Welded construction had totally replaced riveted construction by the 1950s. This 20-year transition period from riveted to welded design and construction was necessary because time was needed to train enough skilled welders and because contractors wanted to keep their skilled riveting crews working as long as possible. Today, design and construction of welded tanks are usually performed under the guidelines of ANSI/AWWA D100. This standard was first published in the November 1935 edition of *Journal - American Water Works Association* as "Standard Specifications for Riveted Steel Tanks and Standpipes" and has undergone several revisions since then.

Advantages

The advent of welded tanks provided opportunities for new tank configurations, but the greatest advantage of welded over riveted tanks was the development of smooth structures with much lower maintenance costs than was possible with lapped, riveted seams. Manual, semiautomatic, and automatic welding processes have improved continually over the years, offering increased economy and strength.

Thicknesses and Capacities

Thicknesses of welded tank shells vary from 3⁄16 in. (4.76 mm) to 2 in. (50 mm) or more. As of the writing of this manual, the largest welded steel water-storage tank constructed had a capacity of 34 mil gal (130 ML). Elevated tanks have been constructed with capacities up to 4 mil gal (15 ML), and designs are available for greater capacities.

BOLTED TANKS

Factory-coated bolted steel storage tanks were developed in the early 1900s to serve as crude-oil and brine containment vessels. In the late 1970s, this tank design gained acceptance for potable water containment through the release of ANSI/AWWA D103, which allows the use of lighter-gauge steel in the production of tank sheets.

Construction

Bolted steel tanks are made of uniformly sized panels (usually 5 ft wide by 8 ft high or 9 ft wide by 5 ft high [1.5 m × 2.4 m or 2.7 m × 1.5 m]), which can be readily transported and assembled at the tank site. Organic gaskets or sealants are used to achieve a watertight seal at the bolted joints. Thicknesses of bolted tank panels vary from a minimum of 0.073 in. (1.85 mm) to 0.500 in. (12.7 mm). Since the panels are bolted together, the tanks can be dismantled and relocated with relative ease.

Capacities

Bolted tanks are currently offered in incremental sizes depending on the tank manufacturer's panel size. Capacities offered range from 4,000 gal (15,000 L) to approximately 2.5 mil gal (6.8 ML). Bolted tanks are available in reservoir and standpipe configurations. Maximum tank heights and capacities are limited by the manufacturer's steel fabrication facility, as well as by ANSI/AWWA D103.

Coating and Life of Tank

Bolted tanks are factory coated for long-term corrosion protection. Four coating systems are presently available for bolted tanks: galvanized, glass, thermoset liquid suspension epoxy, and thermoset powder epoxy. According to the foreword of ANSI/AWWA D103, the anticipated life of a bolted tank is usually limited by the effective life of the protective coating and cathodic protection system. If the coatings are not abused or damaged, the anticipated life expectancy of bolted tanks is more than 30 years.

COMPOSITE ELEVATED TANKS

Composite elevated tanks have been used for water storage since the 1970s, with hundreds of tanks of this style having been constructed and placed into service before the issuance of an AWWA standard. Today the design and construction of composite elevated welded tanks are performed under the guidelines of ANSI/AWWA D107, which was issued in 2010.

NOTE: The scope of ANSI/AWWA D107 is specific to "composite elevated tanks that use a *welded* steel tank for watertight containment" (emphasis added). As of the publication of this edition of Manual M42, no national standard exists that provides guidance on the design, construction, inspection, and testing of composite elevated tanks that use a *bolted* steel tank for watertight containment.

Construction

The composite elevated water tank consists of a concrete support structure (pedestal) and a welded or bolted steel tank. A variety of different styles of pedestal construction are employed by tank fabricators. The top of the pedestal may be constructed with either a concave, convex, or flat top. A welded steel liner is placed over the concrete pedestal top and serves as the floor of the welded steel tank. Bolted tanks may be furnished with a concrete or bolted steel floor.

Part I

Elements of Steel Water Tanks

Typical Capacities and Configurations

Appurtenances

Cathodic Protection

Coating Systems

This page intentionally blank.

Chapter 1

Typical Capacities and Configurations

Steel tanks may be configured as reservoirs, standpipes, or elevated tanks. The configuration selected depends primarily on the required capacity and elevation, as well as on cost. Appearance is also an important consideration.

RESERVOIRS

A reservoir is a ground-supported, flat-bottom cylindrical tank with a shell height less than or equal to its diameter. Reservoirs represent the most common type of water-storage structure. They are used both as a part of the distribution system and to hold treated water for pumping into the distribution system. Of the three types of steel water tanks, a reservoir is generally the most economical to fabricate, erect, and maintain, because of its low height. A photo and a cross-sectional view of a welded steel reservoir are shown in Figure 1-1 and 1-2. A photo and a cross-sectional view of a bolted steel reservoir are shown in Figure 1-3 and 1-4. Table 1-1 gives typical sizes of welded steel reservoirs, and T1-2 gives typical sizes of bolted steel reservoirs and standpipes with different coating systems.

Storage reservoirs for potable water are covered by roof structures, which may be either column supported or self-supporting. Standard tank accessories may include shell and roof manholes, screened roof vents, inside or outside ladders, and connections for pipes as required.

STANDPIPES

Standpipes are ground-supported flat-bottom cylindrical storage tanks that are taller than their diameter. They are usually built where there is little elevated terrain and extra height is needed to create pressure for water distribution. A photo and a cross-sectional view of a welded steel standpipe are shown in Figure 1-5 and 1-6. A photo and a cross-sectional view of a bolted steel standpipe are shown in

Figure 1-1 Welded steel reservoir

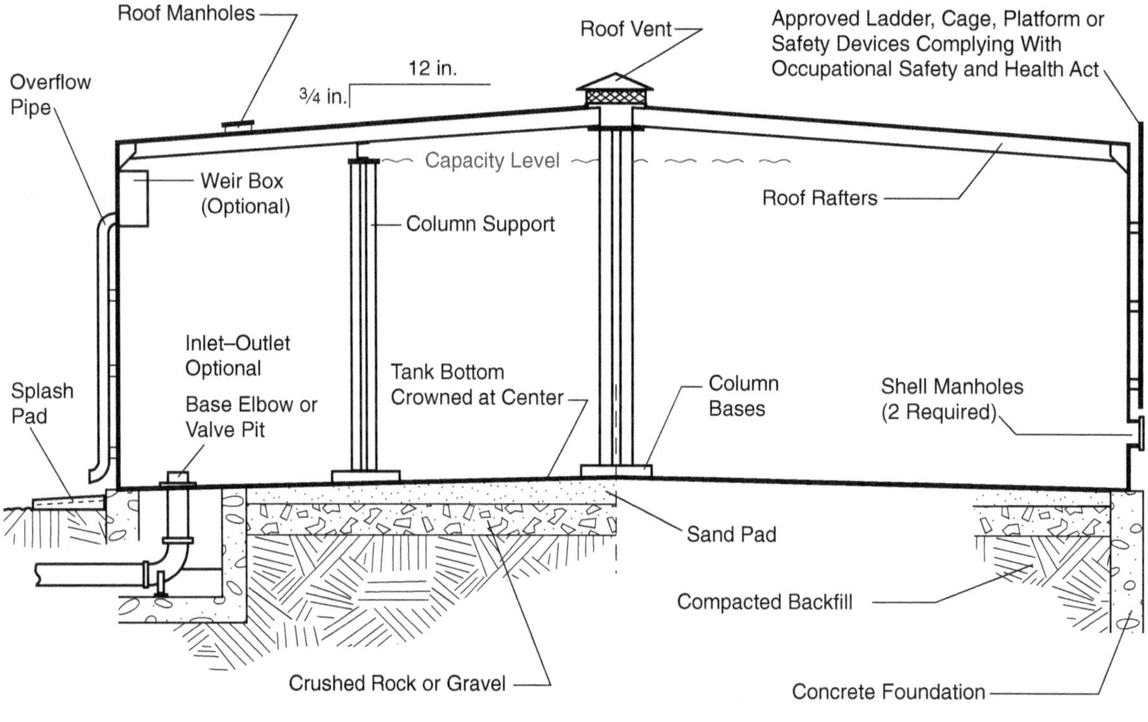

Figure 1-2 Cross-sectional view of a welded steel reservoir

Figure 1-3 Bolted steel reservoir, glass fused to steel

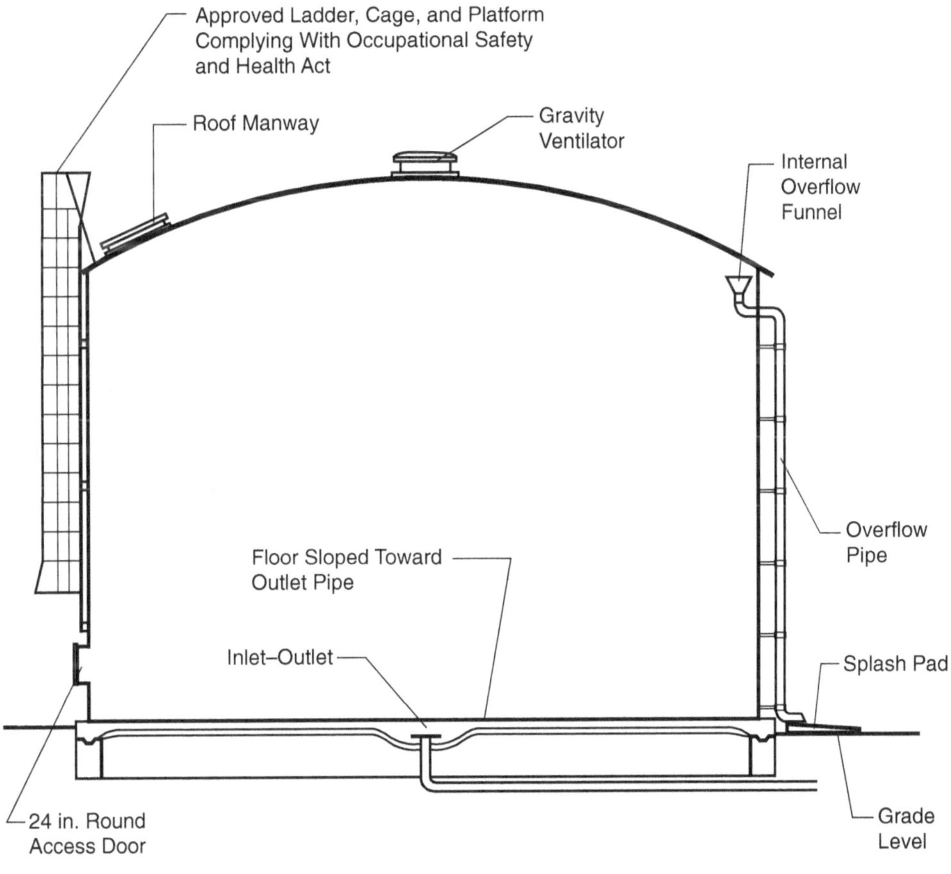

Figure 1-4 Cross-sectional view of a bolted steel reservoir

Table 1-1 Typical welded steel water-storage reservoir sizes

Capacity		Range of Sizes Available			
gal	*m*[3]	Diameter *ft-in.*	Height to TCL *ft-in.*	Diameter *m*	Height to TCL *m*
50,000	189	19-3	24-0	5.9	7.3
60,000	227	21-0	24-0	6.4	7.3
75,000	284	23-6	24-0	7.2	7.3
100,000	379	23-6	32-0	7.2	9.8
		27-0	24-0	8.2	7.3
125,000	473	26-0	32-0	7.9	9.8
		30-3	24-0	9.2	7.3
150,000	568	28-6	32-0	8.7	9.8
		33-0	24-0	10.0	7.3
200,000	757	33-0	32-0	10.0	9.8
		38-3	24-0	11.7	7.3
250,000	946	37-0	32-0	11.3	9.8
		42-9	24-0	13.0	7.3
300,000	1,135	40-6	32-0	12.3	9.8
		46-9	24-0	14.3	7.3
400,000	1,515	46-6	32-0	14.2	9.8
		54-0	24-0	16.5	7.3
500,000	1,890	46-6	40-0	14.2	12.2
		52-0	32-0	15.9	9.8
		60-6	24-0	18.4	7.3
600,000	2,270	51-0	40-0	15.6	12.2
		57-0	32-0	17.4	9.8
750,000	2,840	57-0	40-0	17.4	12.2
		64-0	32-09	19.5	9.8
1,000,000	3,785	66-0	40-0	20.1	12.2
		74-0	32-0	22.6	9.8
1,500,000	5,680	80-6	40-0	24.5	12.2
		90-6	32-0	27.6	9.8
2,000,000	7,570	93-0	40-0	28.4	12.2
		104-6	32-0	31.9	9.8
3,000,000	11,360	114-0	40-0	34.7	12.2
		127-6	32-0	38.9	9.8
4,000,000	15,140	131-6	40-0	40.1	12.2
		147-6	32-0	44.9	9.8
5,000,000	18,930	147-0	40-0	44.8	12.2
		165-0	32-0	50.3	9.8
7,500,000	28,390	180-0	40-0	54.9	12.2
		201-6	32-0	61.4	9.8
10,000,000	37,850	233-0	32-0	71.0	9.8
		208-0	40-0	63.5	12.2

Figures 1-7 and 1-8. Table 1-2 gives typical sizes of bolted steel reservoirs and standpipes. Table 1-3 gives typical sizes of welded steel standpipes.

Operationally, standpipe systems are often designed so that the water in the tank, until it reaches a certain low level, maintains the system pressure. When that low level is reached, pumps come on, valving is changed, and distribution of water is pumped from the lower portion of the standpipe into the system.

Table 1-2 Glass-coated bolted steel reservoirs and standpipes (capacity in thousand gallons*)

Nominal Diameter ft	Nominal Height ft																							
	15	19	24	28	33	38	43	47	52	57	61	66	70	75	79	84	89	93	98	102	107	112	116	121
14	16	22	27	32	37	44	49	54	59	65	70	75	80	86	91	96	101	107	112	117	122	128	133	139
17	24	31	39	47	54	63	70	78	86	93	101	108	116	123	131	139	146	154	161	169	177	184	192	199
20	33	43	53	64	74	86	96	106	117	122	137	148	158	168	179	189	199	210	220	230	241	251	261	272
25	54	71	88	105	122	142	159	176	193	210	227	244	261	278	296	313	330	347	364					
31	81	107	132	158	183	212	238	263	289	320	340	365	391	416	442									
36	114	149	185	220	256	292	327	363	398	434	469	505												
42	151	199	246	294	341	388	436	483	531	578														
50	218	286	355	423	491	559	628	696																
62	326	326	428	530	632	734	836																	
70	421	553	685	816	948																			
81	567	744	921	1,099																				
90	691	906	1,122	1,337																				
101	874	1,147	1,420																					
120	1,247	1,637																						

*To convert feet to meters, multiply by 0.3048; to convert gallons to m^3, multiply by 0.0037854.

As with reservoirs, steel standpipes are covered with a roof structure and may be provided with ornamental trim. Standard accessories may include shell and roof manholes, roof vent(s), a fixed outside ladder, and connections or pipes as required. Inside ladders are not recommended in locations where freezing weather can be expected.

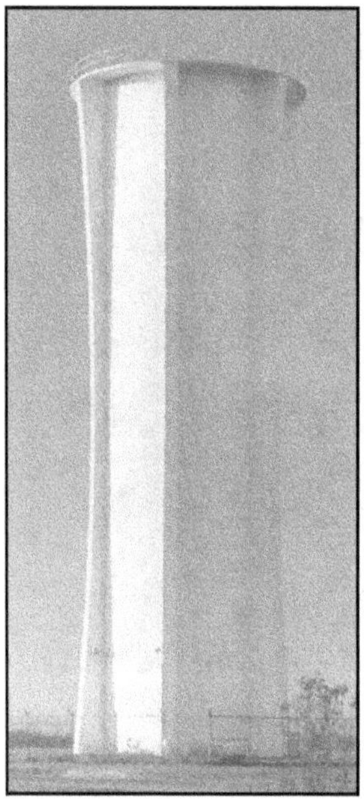

Figure 1-5 Welded steel standpipe with decorative pilasters

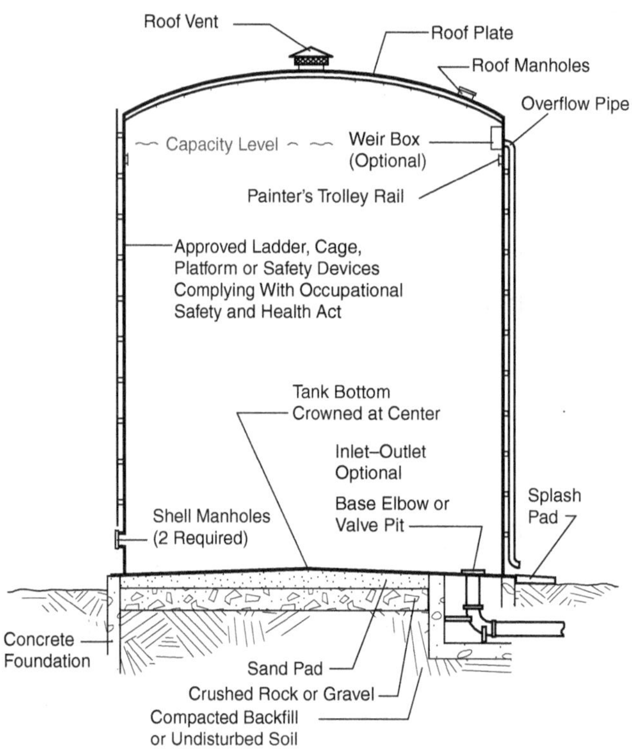

Figure 1-6 Cross-sectional view of a typical welded steel standpipe

Figure 1-7 Bolted steel reservior

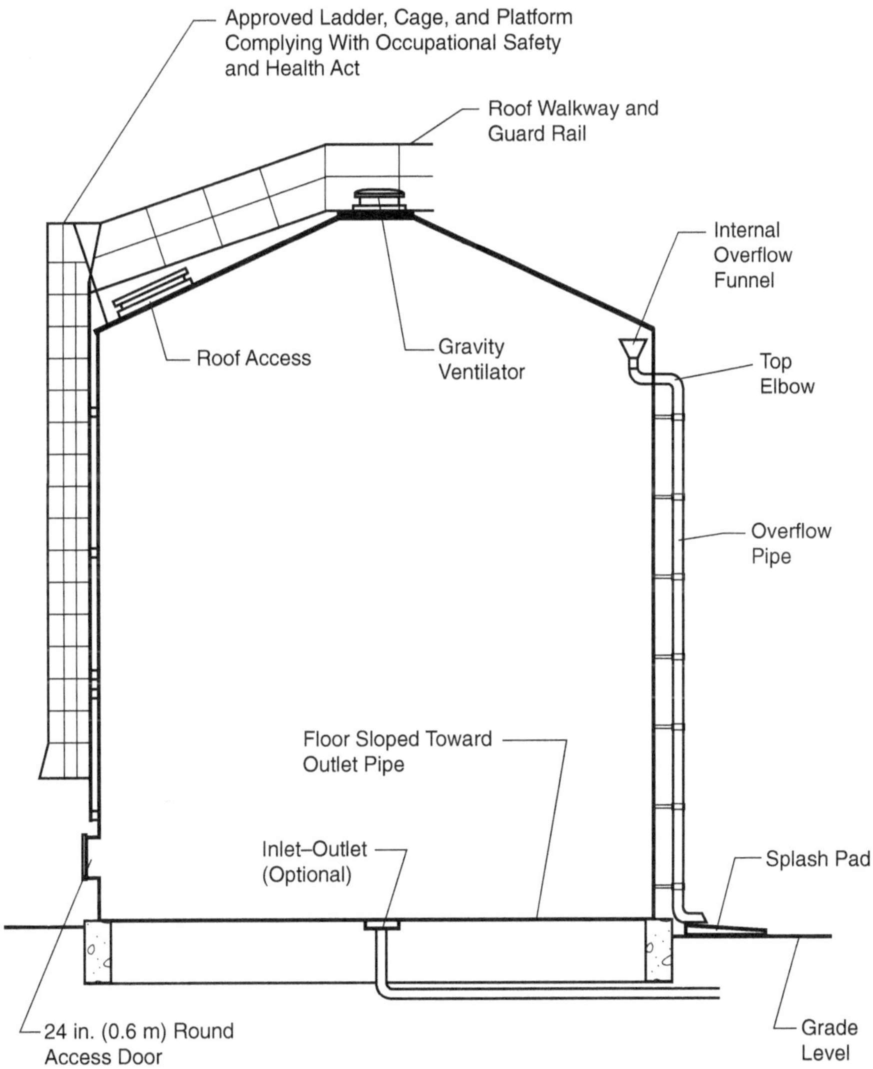

Figure 1-8 Cross-sectional view of a bolted steel standpipe

Table 1-3 Typical welded steel water-storage standpipe sizes

Capacity		Range of Sizes Available			
gal (US)	m^3	Diameter *ft-in.*	Height to TCL *ft-in.*	Diameter *m*	Height to TCL *m*
50,000	189	14-9	40-0	4.5	12.2
60,000	227	16-2	40-0	4.9	12.2
75,000	284	18-0	40-0	5.5	12.2
100,000	379	19-0	48-0	5.8	14.6
125,000	473	21-3	48-0	6.5	14.6
150,000	568	23-3	48-0	7.1	14.6
200,000	757	24-10	56-0	7.6	17.1
250,000	946	27-9	56-0	8.5	17.1
300,000	1,135	28-5	64-0	8.7	19.5
400,000	1,515	32-10	64-0	10.0	19.5
500,000	1,890	34-7	72-0	10.5	21.9
600,000	2,270	37-10	72-0	11.5	21.9
750,000	2,840	42-6	72-0	12.9	21.9
1,000,000	3,785	46-4	80-0	14.1	24.4
1,500,000	5,680	56-9	80-0	17.3	24.4
2,000,000	7,570	65-6	80-0	20.0	24.4
2,500,000	9,465	69-10	88-0	21.3	26.8
3,000,000	11,360	76-6	88-0	23.3	26.8
4,000,000	15,140	84-6	96-0	25.8	29.3
5,000,000	18,930	94-6	96-0	28.8	29.3

ROOF DESIGNS FOR RESERVOIRS AND STANDPIPES

The emphasis on making steel water reservoirs and standpipes attractive as well as functional has led to the development of a wide variety of roof designs. Alternative roof styles for welded tanks include conical, toriconical, umbrella, dome, and ellipsoidal designs, some of which are column supported; others are self-supporting. Bolted steel tanks are usually provided with conical roofs or may be furnished with an aluminum geodesic dome. Column-supported roof structures are not usually employed on steel standpipes greater than 50 ft (15 m) in height. Whichever design is selected, it is particularly important to design any rafters, trusses, columns, stiffeners, and connections so as to minimize potential corrosion sites. All interfaces and connections of such members should be analyzed for their corrosion potential, and protective coatings should be applied to all surfaces deemed necessary from a cost-benefit standpoint.

Column- and Rafter-Supported Cone Roofs

The column- and rafter-supported roof is generally the most economical for a reservoir (Figure 1-9). The roof has a minimum slope for adequate drainage and provides easy access to the manhole for interior inspection. Column loads are spread to a safe limit by column bases, and concrete footings under the columns are not usually required. A modification of this design incorporates a transition from the shell plate to the roof plate that is a smooth curve rather than a sharp break. This transition, or knuckle plate, is a dished or rolled section that usually requires a stiffener at the rafter attachment point (Figure 1-10).

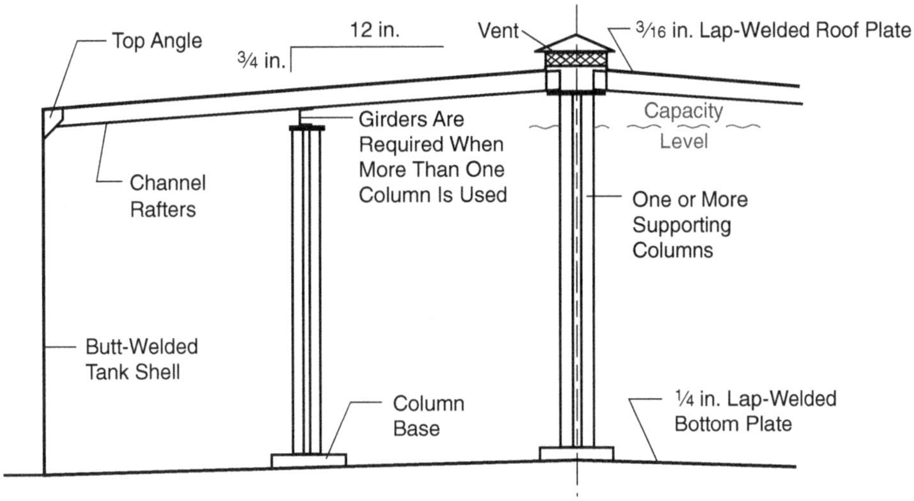

Figure 1-9 Column- and rafter-supported cone roof tank

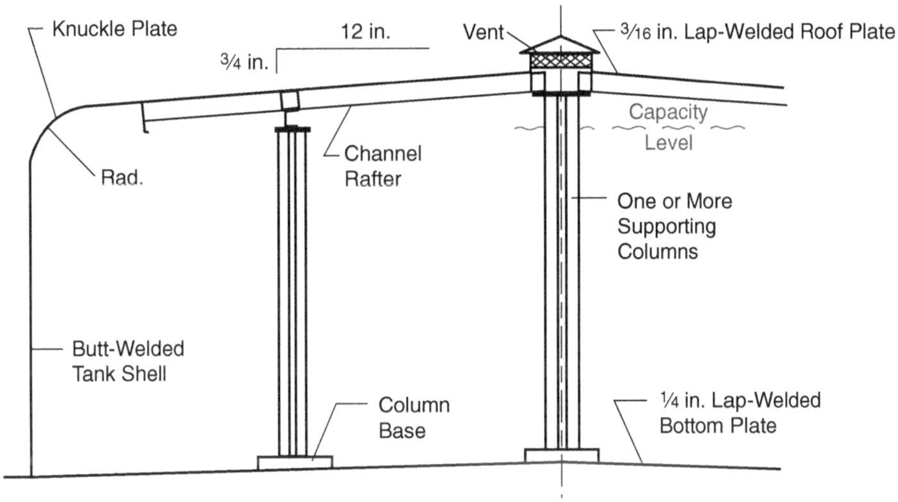

Figure 1-10 Column- and rafter-supported roof with knuckle

Self-Supporting Dome Roof and Umbrella Roof

Steel self-supporting roofs are constructed of plates that are butt welded, lap welded, or lap bolted. They are supported directly on the top angle and shell plate. This type of roof is used where an uncluttered interior and smooth exterior appearance are desired. Dome roof sections are pressed to form a spherical shape. Umbrella roofs are formed to a radius in one direction only, forming chords like the cloth between the spines of an umbrella (Figure 1-11).

Structural stiffeners may be used internally on large-diameter roofs to avoid excessive plate thickness on welded or bolted tanks. Sometimes steel trusses may be used to support the roof, but these should be avoided if possible because they may create corrosion problems and increased coating costs. In addition, the trusses should be kept above the water line to avoid damage by ice and accelerated rates of corrosion.

A modification of the self-supporting dome is the toriconical roof, which consists of a rolled or pressed knuckle and a higher-pitched self-supporting center.

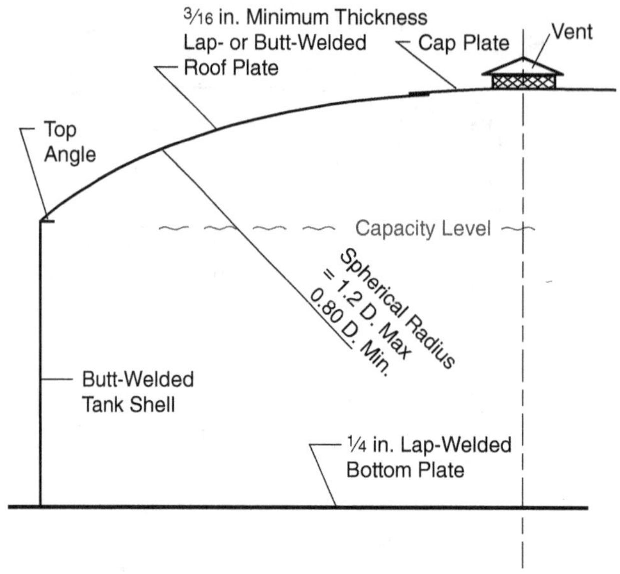

Figure 1-11 Self-supporting dome roof or umbrella roof

Aluminum dome roofs are sometimes erected on bolted or welded steel tanks in accordance with ANSI/AWWA D108, Standard for Aluminum Dome Roofs for Water Storage Facilities. These aluminum domes are usually constructed of triangulated space truss (geodesic) panels. The dead weight of these domes is usually 3 lb/ft^2 (143 N/m^2) or less, as compared with 3.8 lb/ft^2 (181 N/m^2) for a bolted steel roof and 7.6 lb/ft^2 (364 N/m^2) for a welded steel roof.

Self-Supporting Ellipsoidal Roof

The self-supporting ellipsoidal roof is not a true ellipse, but it is formed with two radii yielding major and minor axis proportions of approximately 2:1. The transition from shell to roof is a smooth unbroken curve (Figure 1-12). This roof design is suitable for large- and small-diameter reservoirs and standpipes. On tanks of 50-ft (15-m) diameter or less, the roof is usually free of internal structural members. Larger-diameter tanks usually have radial and circumferential stiffening members or rafters, which may be subject to corrosion problems if they are not properly designed or maintained.

Self-Supporting Cone Roof

An inexpensive and very functional type of roof for small-diameter reservoirs and standpipes is the self-supporting cone roof without internal structural members. This roof will normally be too steep to walk on. Access to manholes and vents by a roof ladder or steps and handrail should be provided. All means of access should be designed individually and installed to comply with current standards.

ELEVATED TANKS

An elevated steel water tank has two primary components: the tank itself and its supporting structure. Such tanks are ordinarily used where ground elevation is insufficient to ensure distribution of water at suitable pressure by gravity. Typically these tanks are of welded construction, but they can also be of bolted construction.

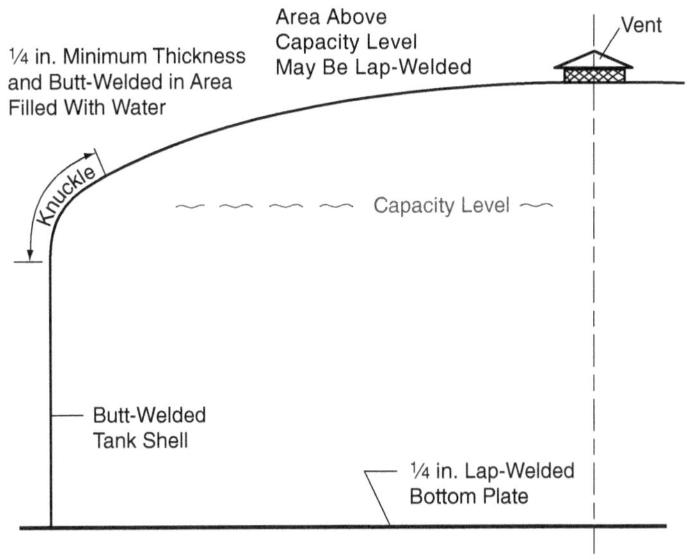

Figure 1-12 Self-supporting ellipsoidal roof

Elevated tanks can be categorized into several different types. The various diameters and head ranges for the tanks shown in the remaining figures and tables in this chapter are only representative and may vary with individual fabricators. Specific diameter/head range combinations should be determined by the tank fabricator within the limits indicated in the tables. Height should be specified by the purchaser as the dimension between the top of the foundation and the top capacity level of the tank. Further dimensions, which are a function of the fabricator's standard, should not be specified. In order to minimize cost, desired operating ranges should be specified to fall within standard available tank dimensions. However, individual operating needs may dictate nonstandard operating ranges.

MULTIPLE-COLUMN ELEVATED TANKS

Small-Capacity Elevated Tanks

The small-capacity multiple-column elevated (or double-ellipsoidal) tank has a cylindrical sidewall, an ellipsoidal bottom and roof, and a top capacity level in the roof several feet above the top of the cylindrical shell. Although in the past they were constructed in capacities up to 1 mil gal (3.8 ML), today double-ellipsoidal tanks are typically constructed only in capacities of 200,000 gal (760,000 L) or less. A photo and a cross-sectional view of a small-capacity elevated tank are shown in Figures 1-13 and 1-14. Table 1-4 gives typical sizes of double-ellipsoidal elevated tanks.

Medium-Capacity Elevated Tanks

For medium-capacity multiple-column elevated tanks, the toroellipsoidal design provides a lower initial cost by most efficiently using the strength of steel. The features used (torus bottom and ellipsoidal roof) cause the central riser to support, as well as contain, a considerable portion of the stored water, while the major portion of the steel bottom acts as a membrane in tension. These tanks usually have a capacity between 200,000 gal (760,000 L) and 500,000 gal (1.9 ML). A photo and a cross-sectional view of a medium-capacity elevated tank are shown in Figures 1-15 and 1-16. Table 1-5 gives typical sizes of medium-capacity elevated tanks.

Figure 1-13 Double-ellipsoidal tank

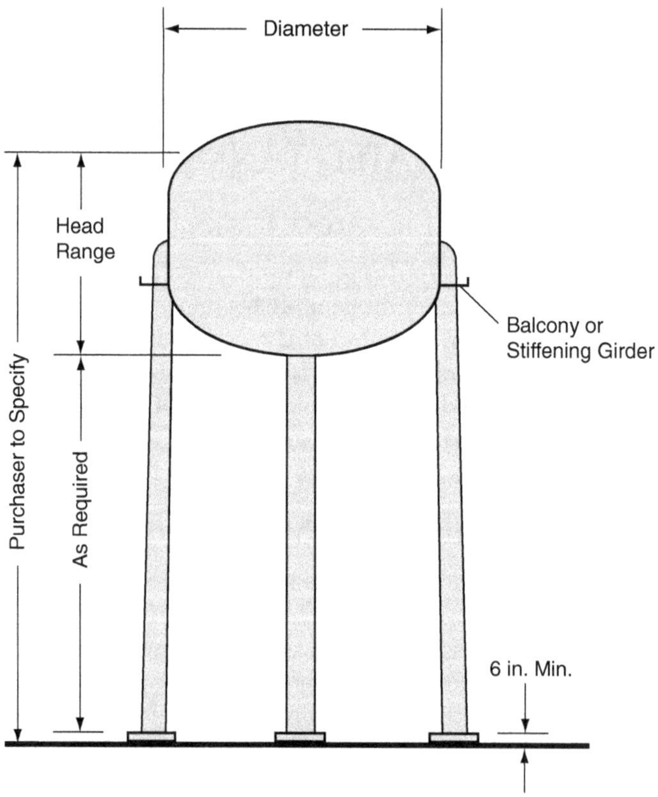

Figure 1-14 Cross-sectional view of double-ellipsoidal tank

Table 1-4 Typical double-ellipsoidal steel elevated tank sizes

Capacity		Range of Sizes Available			
		Diameter	Head Range	Diameter	Head Range
gal	m^3	*ft*	*ft*	*m*	*m*
25,000	95	18–20	12.5–15.5	5.5–6.1	3.3–4.7
30,000	114	18–20	15.0–16.5	5.5–6.1	4.6–5.0
40,000	150	22–23	15.0–17.0	5.7–7.0	4.6–5.2
50,000	189	22–24	18.0–20.0	6.7–7.3	5.5–6.1
60,000	227	22–25	19.0–23.0	6.7–7.6	5.3–7.0
75,000	234	26–30	16.0–24.0	7.9–9.1	4.9–7.3
100,000	379	28–30	20.0–25.0	8.5–9.1	6.1–7.6
125,000	473	30–32	23.0–28.0	9.1–9.7	7.0–8.5
150,000	568	32–34	24.5–29.5	9.7–10.4	7.5–9.0
200,000	757	36–38	28.0–29.5	11.0–11.6	8.5–9.0

Figure 1-15 Medium-capacity welded elevated tank

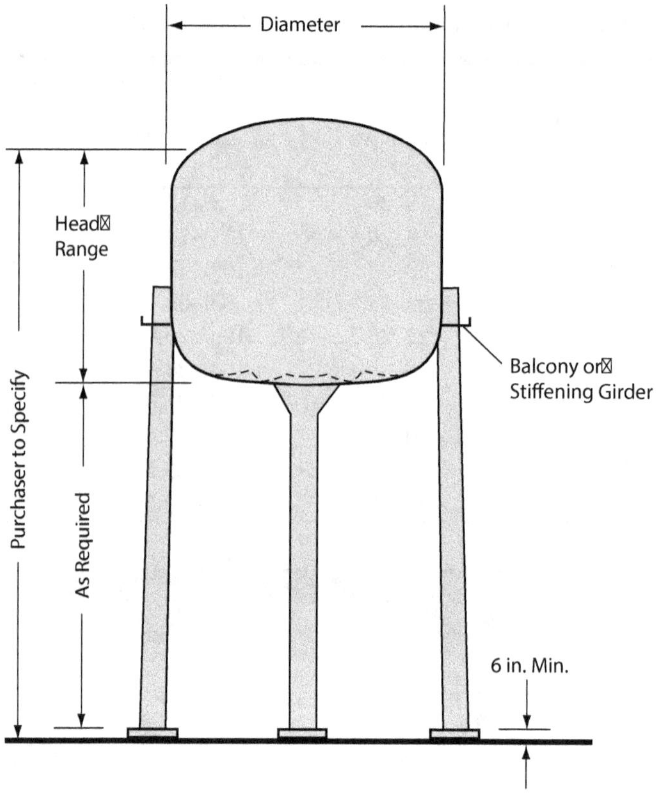

Figure 1-16 Cross-sectional view of medium-capacity torus-bottom welded elevated tank

Table 1-5 Typical medium-capacity welded steel elevated tank sizes

Capacity		Range of Sizes Available			
		Diameter	Head Range	Diameter	Head Range
gal	*m*³	*ft*	*ft*	*m*	*m*
200,000	757	36–38	28–30	11.0–11.6	8.5–9.1
250,000	946	38–40	28–33	11.6–12.2	8.5–10.1
300,000	1,135	43–45	28–31	13.1–13.7	8.5–9.4
400,000	1,515	46–50	30–36	14.0–15.2	9.1–11.0
500,000	1,890	50–56	29–38	15.2–17.1	8.8–11.5
750,000	2,840	56–65	34–45	17.1–19.8	10.4–13.7
1,000,000	3,785	64–65	45–46	19.5–19.8	13.7–4.0

Large-Capacity Multiple-Column Elevated Tanks

Large-capacity elevated tanks (>500,000 gal [1,890 m^3]) provide economical service for communities that need to store a substantial volume of water. Lower operating and pumping costs are ensured because of the low head range, which achieves minimum variation of water pressure throughout the system. A photo and a cross-sectional view of a large-capacity elevated tank are shown in Figures 1-17 and 1-18. Table 1-6 gives typical sizes of large-capacity elevated tanks.

Figure 1-17 Large-capacity elevated tank

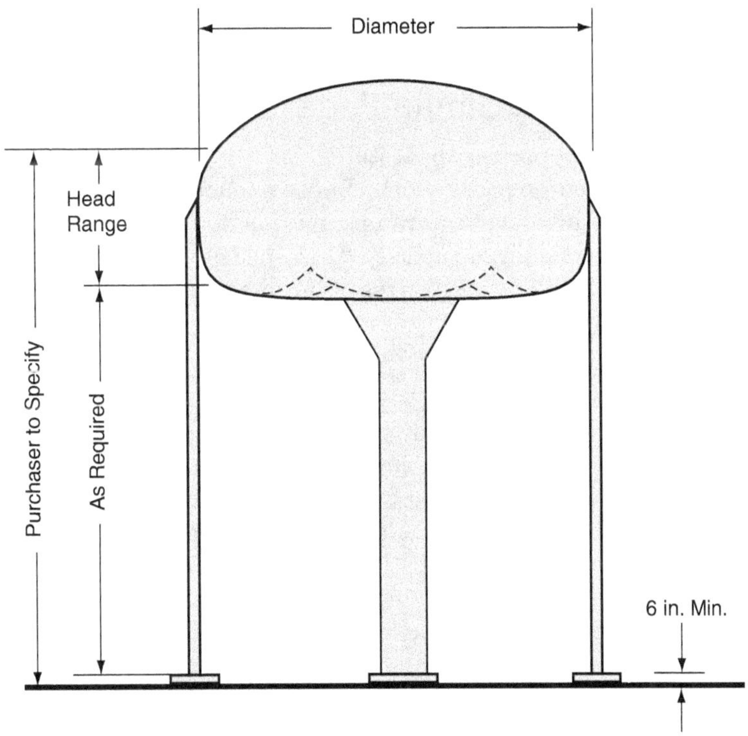

Figure 1-18 Cross-sectional view of large-capacity multi-column elevated tank

Table 1-6 Typical large-capacity welded steel elevated tank sizes

Capacity		Range of Sizes Available			
		Diameter	Head Range	Diameter	Head Range
gal	*m*^{3}	*ft*	*ft*	*m*	*m*
500,000	1,890	60–65	24–25	18.3–19.8	7.3–7.9
600,000	2,270	65–70	24–25	19.8–21.3	7.3–7.9
750,000	2,840	70–76	25–30	21.3–23.2	7.6–9.1
1,000,000	3,735	75–87	25–35	22.9–25.5	7.6–10.7
1,500,000	5,680	91–98	30–35	27.7–29.9	9.1–10.7
2,000,000	7,570	105–106	34–36	32.0–32.3	10.4–11.0
2,500,000	9,465	108–117	39–41	32.9–35.7	11.9–12.5
3,000,000	11,360	119–127	35–40	36.3–38.7	10.7–12.2

PEDESTAL ELEVATED TANKS

Small-Capacity Single-Pedestal Tanks

The single-pedestal spherical tank, widely favored for smaller tank capacities, is often used when appearance is a concern. The gracefully flared base contains sufficient space for pumping units and other operating equipment, a feature common to all pedestal-type vessels.

Ladders to access the container and roof are inside to protect against unauthorized access. Such tanks are usually constructed in capacities of 200,000 gal (760,000 L) or less. A photo and a cross-sectional view of a small-capacity single-pedestal tank are shown in Figures 1-19 and 1-20. Table 1-7 gives typical sizes of small-capacity single-pedestal tanks.

Small-capacity elevated tanks are also constructed as various combinations of cones and cylinders. An alternative design is shown in Figure 1-21.

Large-Capacity Single-Pedestal Tanks

The tubular supporting pedestal gives the large-capacity single-pedestal tank a distinctively contemporary look. Large capacities (0.2–2 mil gal [0.76–7.6 ML]) are provided by this low-head-range spheroidal tank design. A photo and a cross-sectional view of a large-capacity single-pedestal tank are shown in Figures 1-22 and 1-23. Table 1-8 gives typical sizes of large-capacity single-pedestal tanks.

Modified Single-Pedestal Tanks

The attractive modified single-pedestal tank has a central support column, usually fluted to give structural rigidity, that encloses the riser pipe, overflow pipe, and access ladder to the tank roof. The support column may be constructed of steel or concrete. The space within the column can provide multistory usable floor space for pumping, storage, and office facilities. Although available in all capacities, these tanks are not usually constructed in capacities below 500,000 gal (1.9 ML). A photo and a cross-sectional view of a modified single-pedestal tank are shown in Figures 1-24 and 1-25. Table 1-9 gives typical sizes of modified single-pedestal tanks.

Figure 1-19 Spherical single-pedestal tanks give pleasant silhouette

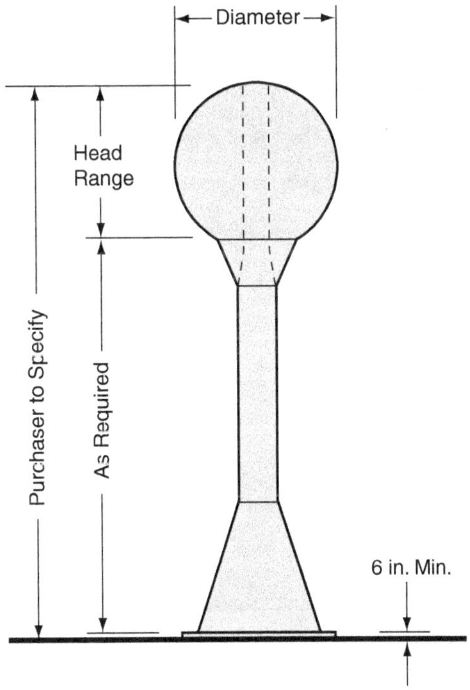

Figure 1-20 Cross-sectional view of small-capacity spherical single-pedestal tank

Table 1-7 Typical small-capacity single-pedestal steel elevated tank sizes

Capacity		Range of Sizes Available			
		Diameter	Head Range	Diameter	Head Range
gal	*m^3*	*ft*	*ft*	*m*	*m*
25,000	95	19–20	15–17	5.8–6.1	4.6–5.2
30,000	114	20–21	15–18	6.1–6.4	4.6–5.5
40,000	150	21–23	19–22	6.4–7.0	5.8–6.7
50,000	189	23–24	19–23	7.0–7.3	5.8–7.0
60,000	227	24–26	22–24	7.3–7.9	6.7–7.3
75,000	234	25–28	23–27	7.9–8.5	7.0–8.2
100,000	379	29–30	25–30	8.8–9.1	7.6–9.1
125,000	473	31–33	27–32	9.4–10.0	8.2–9.7
150,000	568	33–34	30–34	10.1–10.4	9.1–10.4
200,000	757	36–38	36–38	11.0–11.6	11.0–11.6

Figure 1-21 Alternative single-pedestal tank design

Figure 1-22 Large-capacity single-pedestal elevated tank

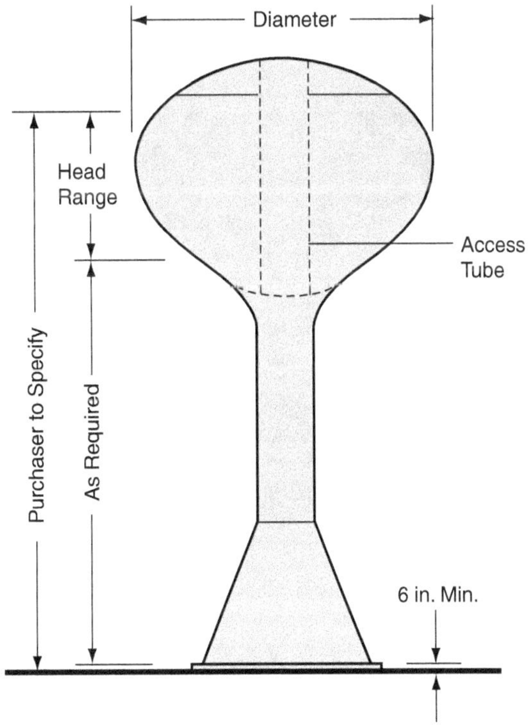

Figure 1-23 Cross-sectional view of large-capacity single-pedestal elevated tank

Table 1-8 Typical large-capacity single-pedestal steel elevated tank sizes

Capacity		Range of Sizes Available			
gal	m^3	Diameter *ft*	Head Range *ft*	Diameter *m*	Head Range *m*
200,000	757	40–42	27–30	12.2–12.8	8.2–9.1
250,000	946	43–47	25–32	13.1–14.3	7.6–9.7
300,000	1,135	46–48	30–33	14.0–14.6	9.1–10.1
400,000	1,515	50–53	30–40	15.2–16.1	9.1–12.2
500,000	1,890	55–60	30–40	16.3–18.3	9.1–12.2
750,000	2,840	64–66	38–42	19.5–20.1	11.6–12.3
1,000,000	3,785	74–78	35–40	22.5–23.8	10.7–12.2
1,250,000	4,750	76–80	40–45	22.9–24.4	12.2–13.7
1,500,000	5,880	85–90	45–50	25.9–27.4	13.7–15.2
2,000,000	7,570	90–95	50–55	27.4–29.0	15.2–16.3

Figure 1-24 Folded-plate design of a modified single-pedestal tank support

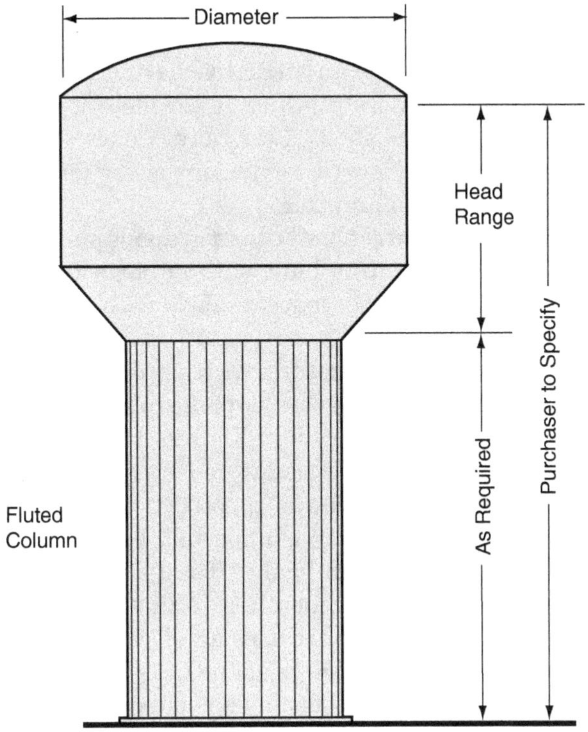

Figure 1-25 Cross-sectional view of modified single-pedestal tank

Table 1-9 Typical modified single-pedestal steel elevated tank sizes

Capacity		Range of Sizes Available			
		Diameter	Head Range	Diameter	Head Range
gal	m^3	*ft*	*ft*	*m*	*m*
250,000	946	41–43	29–31	12.5–13.1	8.8–9.4
300,000	1,135	43–45	29–31	13.1–13.7	8.8–9.4
500,000	1,890	49–64	30–39	14.9–19.5	9.1–11.9
750,000	2,840	63–65	37–40	19.2–19.8	11.3–12.2
1,000,000	3,785	73–78	35–42	22.2–23.8	10.7–12.8
1,500,000	5,680	85–87	39–46	25.9–26.5	11.9–14.0
2,000,000	7,570	97–102	38–46	29.6–31.1	11.6–14.0
2,500,000	9,465	107–110	43–45	32.6–33.5	13.1–13.7
3,000,000	11,360	109–120	40–45	33.3–36.6	12.2–13.7

Composite Elevated Tanks

Composite elevated tanks are an attractive design that uses the best design features of steel and concrete. Concrete, which is excellent for compression loads, is used as the support column for the steel water tank. Concrete has the advantage of requiring either no painting or a low cost exterior coating for aesthetic purposes. Two tank types are available: welded and bolted.

Welded tanks. The steel bowl construction is similar to that found on the fluted-column tanks; the bowl can be built with either a cone or a domed roof. The most common designs use a domed concrete floor with a steel liner. Commonly built to store between 500,000 gal and 2 mil gal (1.7 and 7.6 ML), these tanks provide many of the benefits of a fluted-column tank with significantly less area requiring painting, thereby reducing maintenance costs. A photo and a cross-sectional view of a composite elevated welded tank are shown in Figures 1-26 and 1-27. Table 1-10 gives typical sizes of composite elevated welded tanks.

Bolted tanks. The design and construction of an elevated bolted steel tank is similar to those used for a ground-supported, bolted steel tank. Common capacities range from 150,000 gal to 1.5 mil gal (0.6 and 5.7 ML). The tanks are factory-coated and supported by a reinforced concrete slab that is constructed at the top of the concrete support column. The slab becomes an elevated foundation for the tank and can be designed and constructed to serve as the tank bottom, or a bolted steel bottom may be erected on top of the slab. Roofs may be bolted steel or bolted aluminum domes. A photo and a cross-sectional view of a composite elevated bolted tank are shown in Figures 1-28 and 1-29. Table 1-11 gives typical sizes of composite elevated bolted tanks.

Courtesy of Walter Lagarenne

Figure 1-26 Composite elevated welded tank

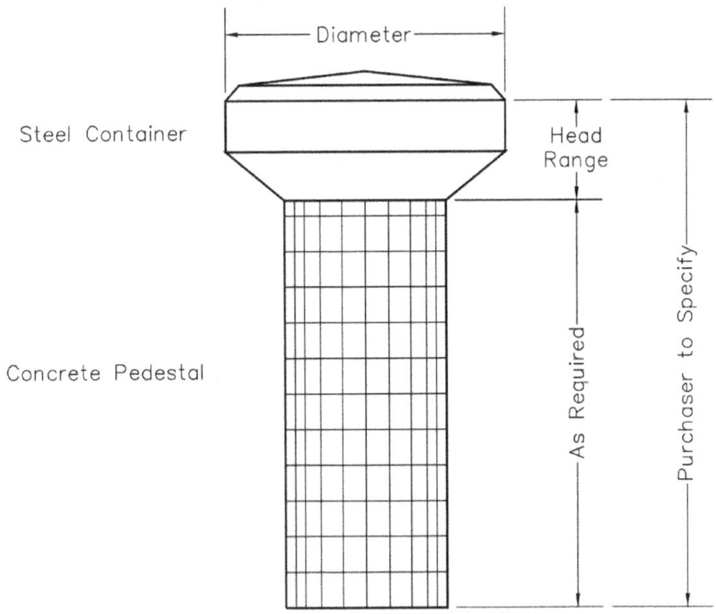

Courtesy of Tank Industry Consultants

Figure 1-27 Cross-sectional view of a composite elevated welded tank

Table 1-10 Typical composite elevated welded tank sizes

Capacity		Range of Sizes Available			
		Diameter	Head Range	Diameter	Head Range
gal	m^3	*ft*	*ft*	*m*	*m*
500,000	1,890	50–55	35–40	16.3–18.3	10.7–12.2
750,000	2,840	58–60	38–42	19.5–20.1	11.6–12.8
1,000,000	3,785	68–72	38–42	22.5–23.8	11.6–12.8
1,250,000	4,750	78–80	38–42	23.8–24.4	11.6–12.8
1,500,000	5,680	85–90	38–42	25.9–27.4	11.6–12.8
2,000,000	7,570	90–95	42–47	27.4–29.0	12.8–14.3
2,500,000	9,460	103–107	42–47	31.4–32.6	12.8–14.3
3,000,000	11,360	115–120	42–47	35.1–36.6	12.8–14.3
3,500,000	13,250	115–120	48–52	35.1–36.6	14.6–15.8

Courtesy of CST Storage

Figure 1-28 Composite elevated bolted tank

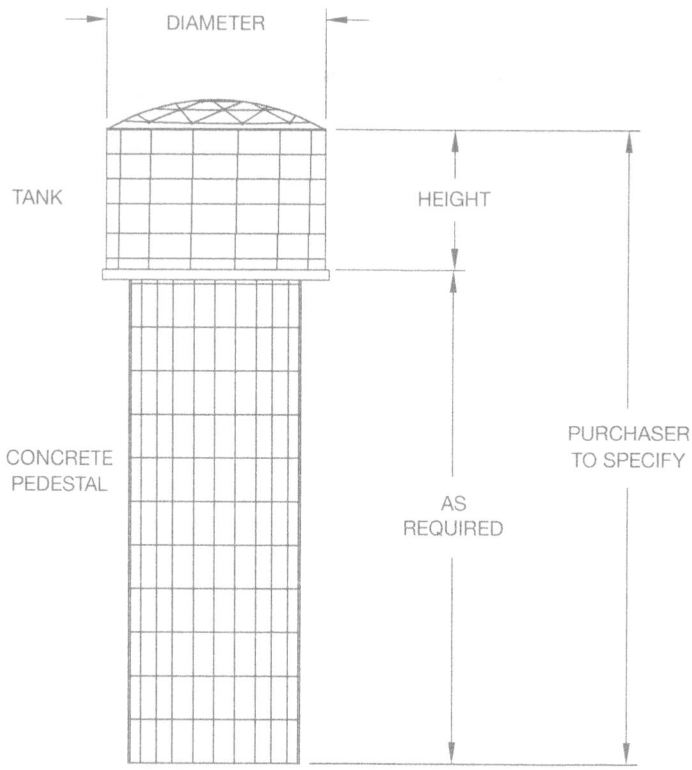

Courtesy of CST Storage

Figure 1-29 Cross-sectional view of a composite elevated bolted tank

Table 1-11 Typical composite elevated bolted tank sizes

Capacity		Range of Sizes Available			
gal	*m³*	Diameter *ft*	Height *ft*	Diameter *m*	Height *m*
150,000	568	28	33	8.5	10.1
200,000	757	31	33	9.4	10.1
300,000	1,140	39	33	11.9	10.1
500,000	1,890	50	33	15.2	10.1
750,000	2,840	59	38	18.0	11.6
1,000,000	3,785	62	47	18.9	14.3
1,500,000	5,680	67	56	20.4	17.1

This page intentionally blank.

Chapter 2

Appurtenances

This chapter briefly describes the construction and function of common appurtenances for steel water-storage tanks. Appurtenances are covered in separate sections of ANSI/AWWA D100, Standard for Welded Carbon Steel Tanks for Water Storage, ANSI/AWWA D103, Standard for Factory-Coated Bolted Carbon Steel Tanks for Water Storage, and ANSI/AWWA D107, Standard for Composite Elevated Tanks for Water Storage. The majority of the appurtenances and attachments described for steel water-storage tanks are required by law, code, and industry standards to make the tank a safe and functional facility. Other accessories are optional and may be specified by the owner to improve the facility's function or appearance.

SHELL MANHOLES

A minimum of two opposing shell manholes are required on welded ground-supported tanks for ventilation during interior coating operations, for safety, and for ease of interior access during construction activities and maintenance inspections. On tanks more than 100 ft (30.5 m) in diameter, it may be desirable to have three or more shell manholes. ANSI/AWWA D103 requires only one shell manhole on bolted tanks because a tank panel can be removed to provide additional ventilation.

Sizes and Types

Normal manhole sizing is 24 or 30 in. (610 or 760 mm) in diameter to accommodate ventilating equipment and allow easy egress (Figures 2-1 and 2-2). Outward-opening, hinged covers or single-bolt, inward-opening manholes are standard. Outward-opening covers may require reinforcing plates on the shell, whereas inward-opening manholes usually accomplish their reinforcement through heavy plate necks. The heavy plate neck also provides the gasket surface to the cover. The inward-opening cover must be hinged to ensure proper operation. If a tank will be subject to severe icing conditions, an inward-opening manhole may not be desirable.

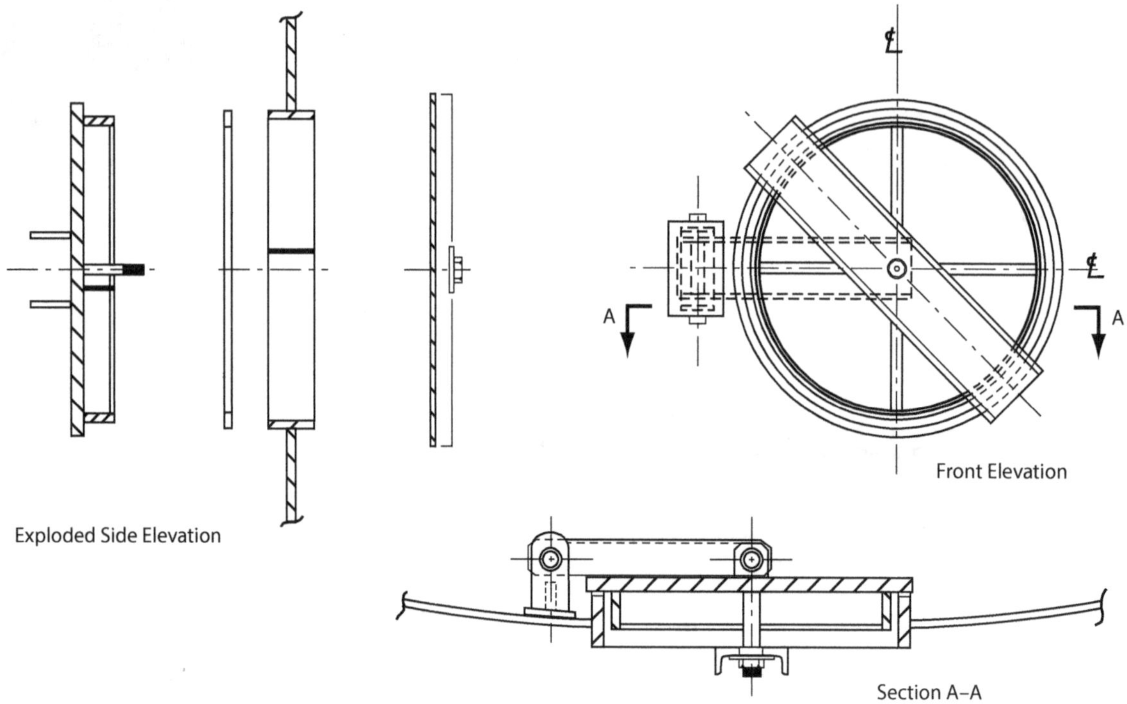

Figure 2-1 Inward-opening shell manhole detail

Bolted Tank Manholes

The shell manholes for bolted tanks can be circular, 24 in. (610 mm) or 30 in. (760 mm) in diameter; square, 24 by 24 in. (610 by 610 mm); or elliptical, 18 by 22 in. (460 by 560 mm), minimum size. One manhole, unless otherwise specified, shall be provided in the first ring of the tank at a location designated by the purchaser. If the manhole cover weighs more than 50 lb (23 kg), a hinged cover should be provided.

Flush Manholes

For welded steel flat-bottom tanks (ANSI/AWWA D100), flush rectangular manholes (rectangular manholes mounted flush with the bottom of the tank) having a minimum length of 24 in. (610 mm) in the shortest direction and a maximum length of 48 in. (1,220 mm) in the longest direction are also available. Such manholes are useful when a tank interior is being cleaned. Refer to ANSI/AWWA D100 and American Petroleum Institute Standard API 650 for details and design requirements.

PIPE CONNECTIONS

The number of tank-bottom or shell-piping connections should be kept to a minimum. The usual practice is to use a common inlet-outlet drain connection through the tank bottom or on the tank shell (Figures 2-3 and 2-4). If a bottom connection is used, a removable section of pipe 6–8 in. (150–200 mm) long may extend above the drain at floor level to serve as a silt stop. Recent requirements concerning minimum and maximum detention time that the water remains in the tank may require separate inlet and outlet connections. Baffles and flow diverters are also used to control tank water detention time.

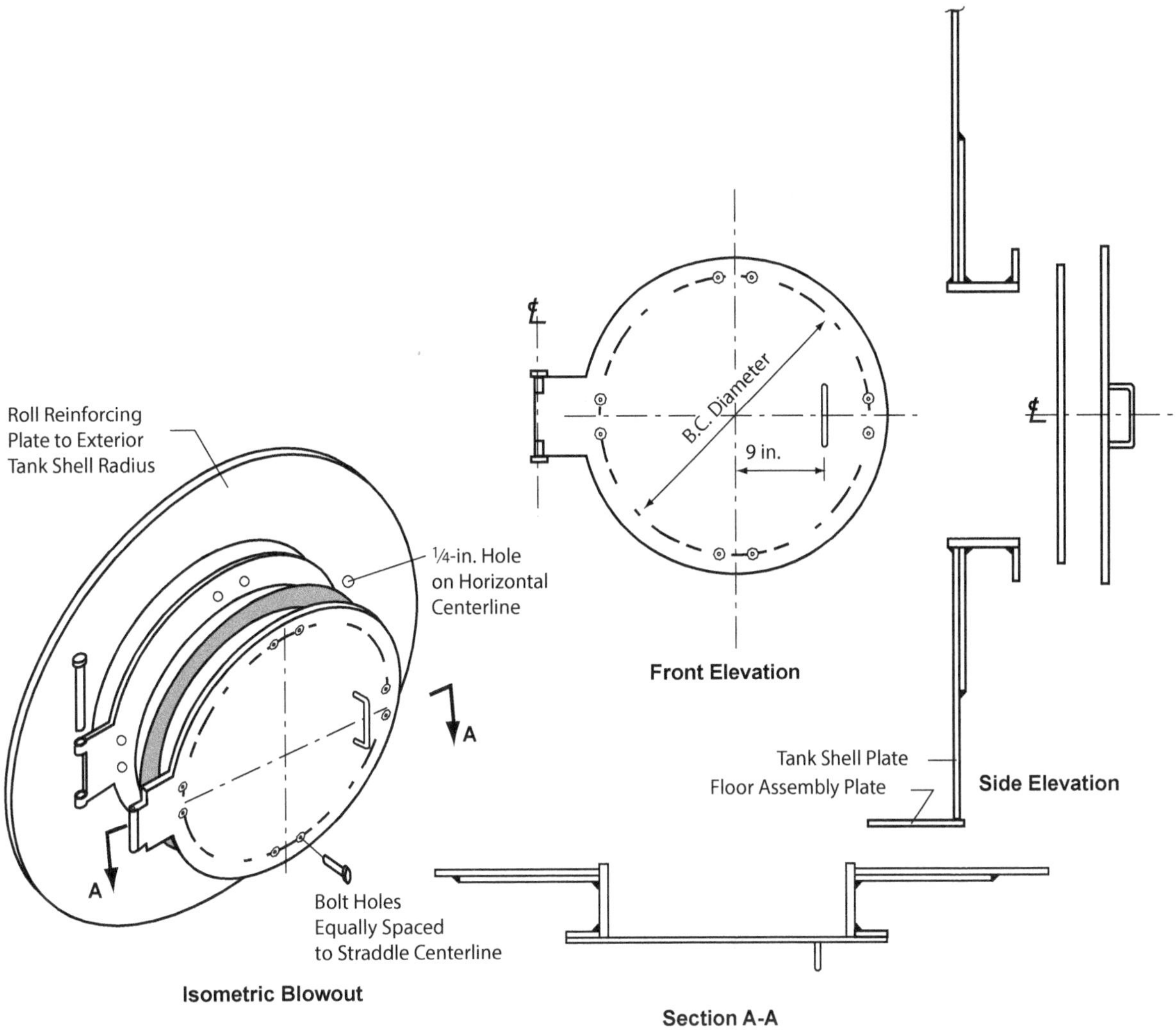

Figure 2-2 Outward-opening shell manhole detail

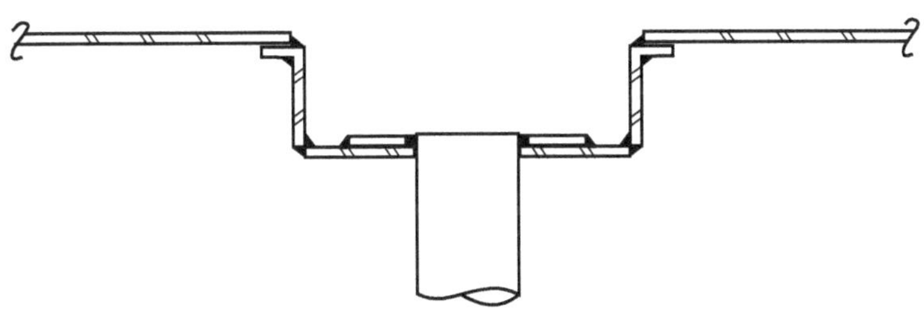

Figure 2-3 Recessed inlet-outlet pipe bottom connection detail

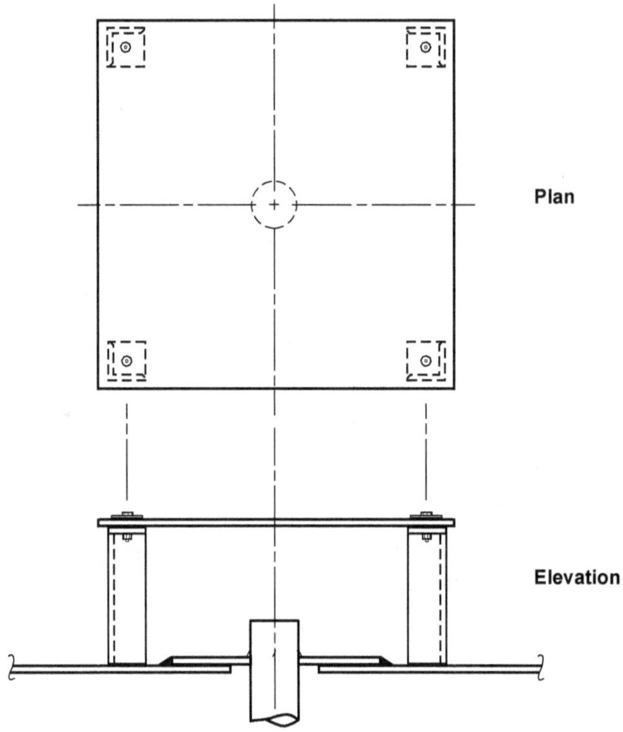

Figure 2-4 Non-recessed inlet-outlet pipe bottom connection details

Piping connections through the tank bottom or shell are normally furnished in steel pipe, welded or bolted to the shell or bottom. Ductile-iron or cast-iron pipe connections must pass through a mechanical-joint type connection that is welded or bolted to the steel tank bottom.

The piping connections to the tank should be constructed to have a flexible coupling outside the tank to allow for differential movement. If piping connections through the shell are required, provisions must be made to protect this piping from freezing and vandalism. Adequate precautions must be made in this piping to allow for differential movement when the tank is filled and drained. Special flexible, extendable connections are necessary for tanks subject to seismic movement. ANSI/AWWA D100 defines the distance from the shell intersection where through-the-tank-bottom piping connections may be located on unanchored tanks designed for seismic conditions.

OVERFLOW

A properly sized overflow is essential to protect the tank structure from excessive water levels caused by rapid variations in distribution system conditions. Exterior overflows are recommended. In colder climates, ice buildup on an internal overflow may become a problem and eventually break the overflow pipe. Overflow waters should be directed beyond the tank exterior perimeter to avoid damage to the tank grade or foundation during overflow. Most state standards recommend that the overflow on elevated tanks be extended down the side of the tank to within approximately 12 to 24 in. (305 to 610 mm) above grade. Extending the overflow prevents water discharged from the pipe from freezing on the tower structure and causing damage to the structure. In addition, most governing agencies require an air gap between the overflow tank piping and final drainage system in order to protect

Figure 2-5 Overflow air break with flap valve

against backflow. Figure 2-5 shows one type of overflow-pipe air gap. Most states require a screening, elastomeric check valve, or flap-gate arrangement over the end of the pipe connected to the tank. As distribution systems and pumping capacities are increased, the vent and overflow capacities of existing tanks should be reevaluated to ensure their adequacy to relieve potential pressure or vacuum conditions in the tank. Overflows should be easily accessible for maintenance, repair, and inspection.

LADDERS AND SAFETY DEVICES

Safe access must be provided for authorized personnel who need to reach the top of the tank facility.

Exterior Ladders

Exterior ladders, cages, and platforms designed to meet Occupational Safety and Health Administration (OSHA) standards are recommended (Figure 2-6). Either the ladder should terminate at least 8 ft (2.4 m) above grade or a solid locking door, provided to discourage unauthorized access to the tank, should be installed on the lower 8–20 ft (2.4–6.1 m) of the exterior ladder. Certain areas will require a locking door and anti-climb screening at the bottom of the ladder cage to discourage unauthorized access. The exterior ladder, roof hatch opening, and interior ladder (if specified) should be located close together to reduce the movement necessary by a climber on the tank roof.

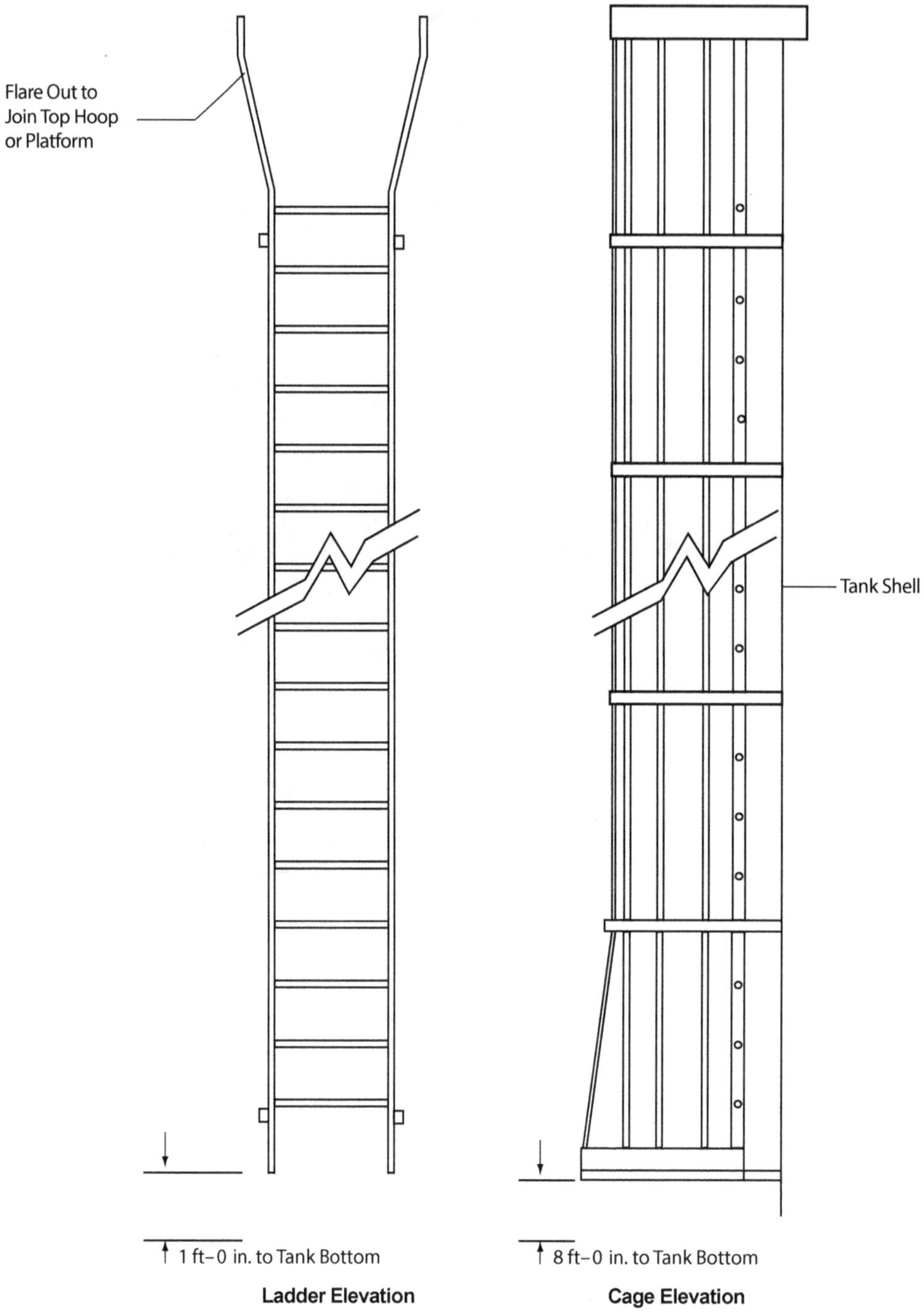

NOTE: All ladders should be designed per OSHA requirements.

Figure 2-6 Exterior caged ladder details

Figure 2-7 Safe-climbing rail for an outside ladder

Interior Ladders

Because of accelerated rates of corrosion and the potential for ice buildup in areas where freezing temperatures occur, ladders inside the tank container are not recommended. Ice buildup on an interior ladder can impose loads on the tank wall plates that are sufficient to pierce or rupture the tank container. Even in temperate climates, corrosion can damage interior ladders, making them unsafe. The use of stainless steel ladders is not recommended as the presence of the dissimilar metal can cause severe corrosion of the carbon steel. However, if used, such ladders must include insulation (dielectric connections) of the stainless steel from the carbon-steel tank and coating of all stainless-steel components or corrosion will occur on the carbon-steel tank.

Ladders are installed inside dry risers and access tubes. There they are not subjected to corrosive conditions, and the access doors may be locked to deter access.

Safety Devices

For tanks with a total height in excess of 20 ft (6.1 m) or as required by OSHA, ladders with offset rest platforms (secondary platforms) are required every 20 ft (6.1 m) or as required by OSHA for the ladder assembly. This length may be extended to 30 ft (9.1 m) if an approved safety cage is used. In lieu of intermediate platforms, approved unpainted safe-climbing rails or cables may be used. Figure 2-7 shows a typical safe-climbing rail.

In addition to exterior ladder safety devices, the most commonly installed safety items are safety railings at the roof where the exterior ladder terminates (Figure 2-8). These railings protect personnel on the roof near the roof hatch. All safety railings installed on the tops of tanks and ladders should comply with minimum OSHA requirements or local building codes. Consult with the applicable agency for tank location to determine the latest safety requirements. Total perimeter handrails are not recommended in high snow-load areas.

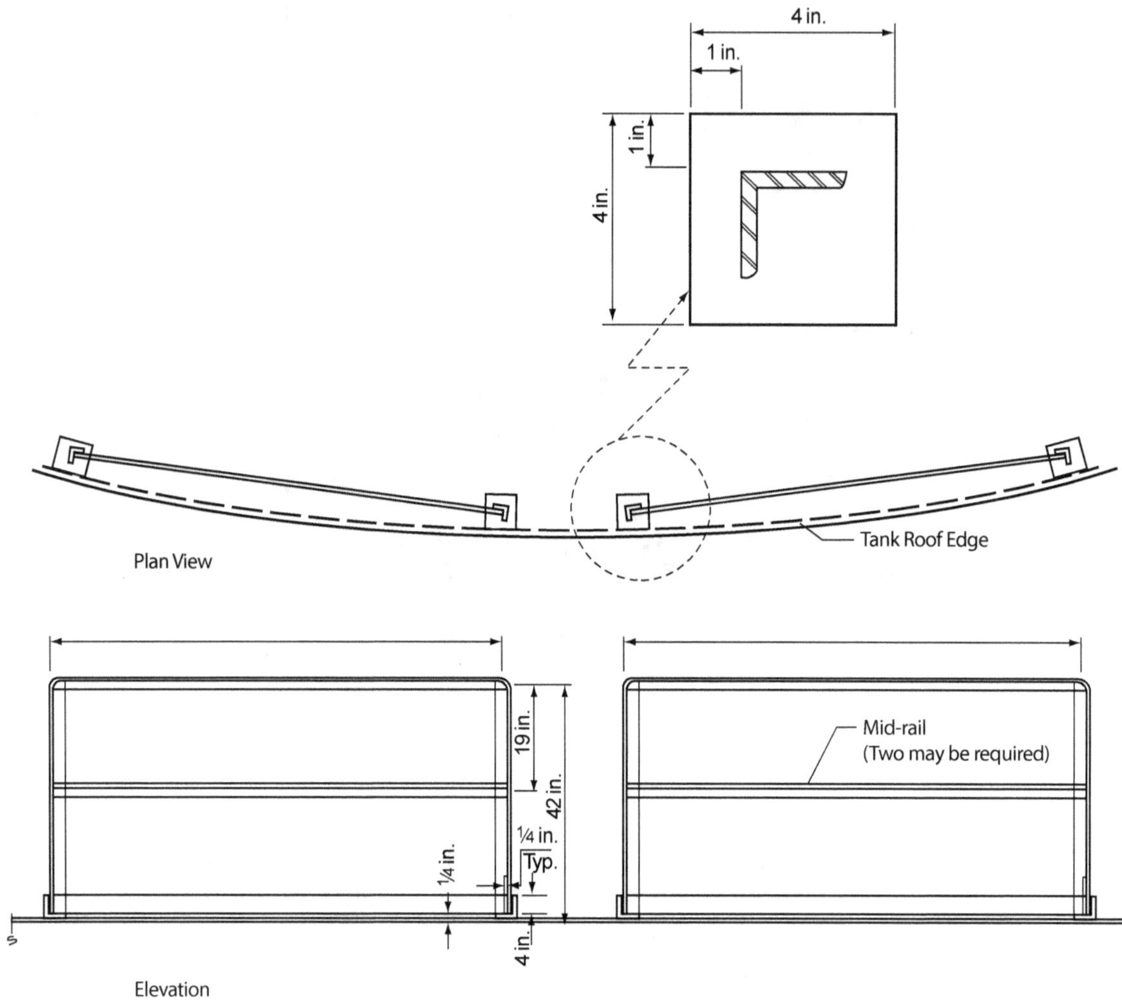

Figure 2-8 Roof guardrail details

ROOF OPENINGS

A minimum of two roof openings is required for personnel access and ventilation during maintenance and rehabilitation activities on welded steel tanks.

Primary Opening

The first (primary) roof opening should be located near the tank sidewall, close to the exterior ladder. The minimum size for this roof opening is 15 by 24 in. (380 by 610 mm), but it is recommended that either a 24-in. (610-mm) or 30-in. (760-mm) square or round opening, with a hinged cover and locking hasp, be used to facilitate access to the tank interior. A minimum 4-in. (100-mm) high curb and a 2-in. (50-mm) downward cover overlap are mandatory on any roof opening to prevent rain or snowmelt from entering the tank (Figure 2-9). Bolted and gasketed roof manways without the curb and overlap are allowed on bolted tanks.

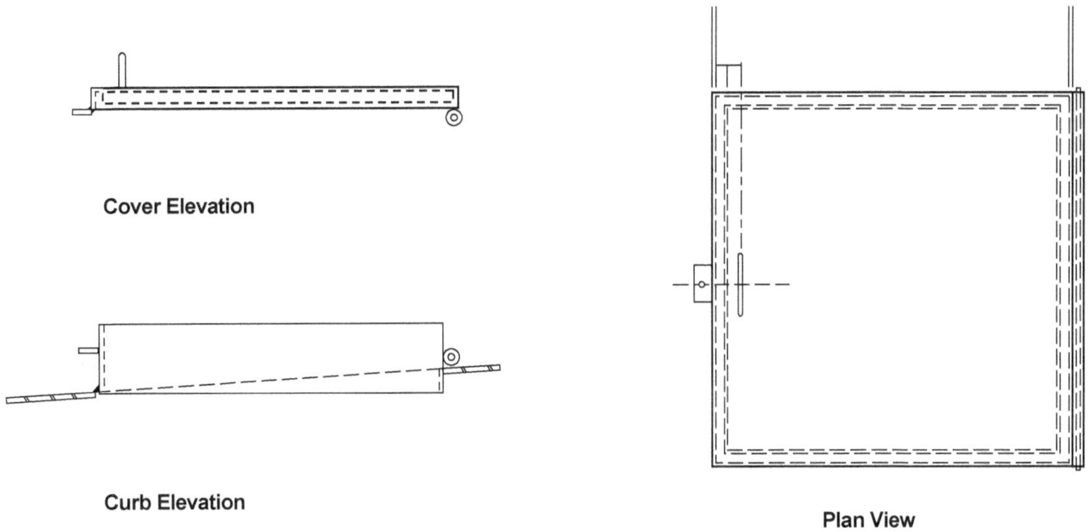

Figure 2-9 Roof manhole assembly details

Secondary Opening

The second roof opening should be located approximately between the tank center and the sidewall directly opposite from the primary opening. It should have at least a 24-in. (610-mm) diameter. This opening may be a standard bolt-cover manhole; however, if the center vent is of adequate size, is not obstructed, and has a removal cover, the vent itself may suffice as the secondary opening. This opening should be designed to accommodate a ventilation fan to be bolted on. All removable covers should be secured to the roof of the tank using hinges or chains. An additional safety rail may be required between the secondary roof openings and the edge of the roof.

VENTS

For closed-top tanks, venting must be provided to safeguard against excess pressure or vacuum buildup during the maximum inflow or outflow of water (Figure 2-10). Structural failures of tanks can be caused by inadequate venting. When the vents are being sized, the area contributed by the overflow pipe should not be considered as part of the ventilation area. A minimum of one vent is required; this should be located near the center of the roof. For larger-diameter tanks, several vents could be located around the other periphery as well as at the center of the tank to facilitate cross-flow ventilation. Chapter 8 discusses vent screening and protection from insects, birds, and other intrusions.

The most common forms of tank vents are the mushroom, pan (Figure 2-11), and 180° types. Vents with pressure- and vacuum-releasing pallets are recommended. An example of a clog-resistant vent is shown in Figure 2-12. All vents should be screened to protect against the entry of birds, animals, and insects. The screening should be stainless steel or some other type of corrosion-resistant material. Shields should be installed to keep dirt and debris from blowing into the tank. In areas of snow buildup, the vents should be protected or elevated to prevent their being clogged by snow.

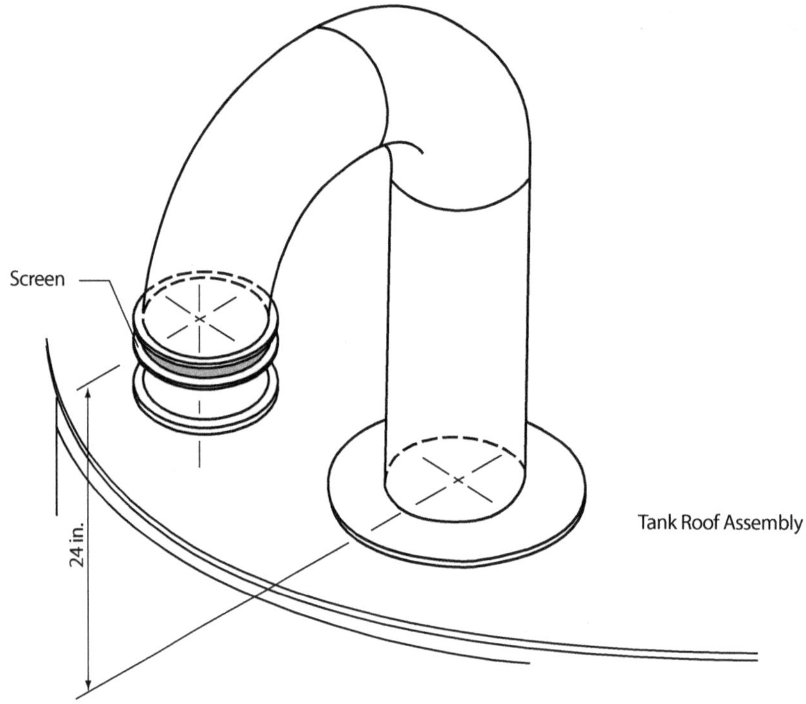

Figure 2-10 Double 90° elbow roof vent detail

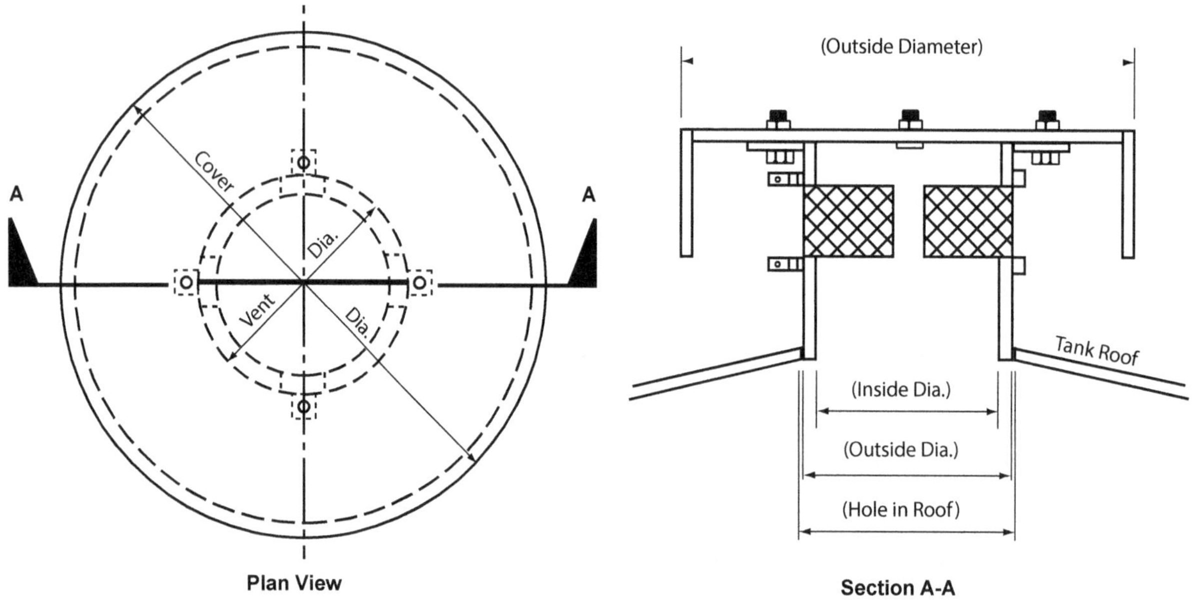

Figure 2-11 Pan deck vent detail

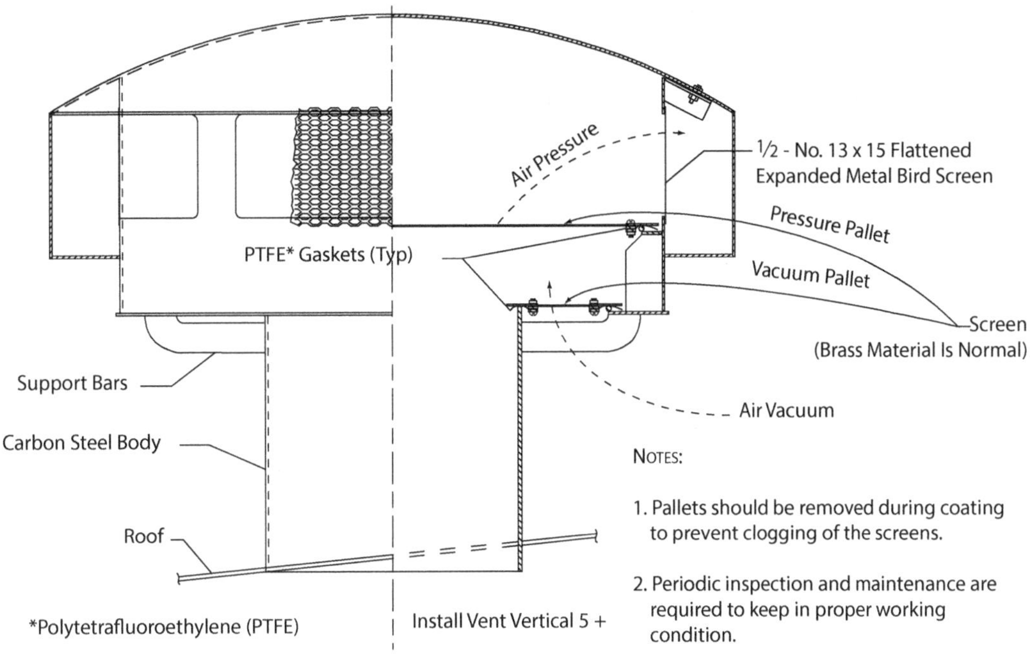

NOTES:

1. Pallets should be removed during coating to prevent clogging of the screens.
2. Periodic inspection and maintenance are required to keep in proper working condition.

Figure 2-12 Typical clog-resistant vent detail

Many older riveted tanks do not have vents; instead they have finial balls that provide limited or no ventilation area. These finial balls should be replaced with vents when maintenance or repair work is done on the tank. As distribution systems and pumping capacities are enlarged, the vent and overflow capacities on existing tanks should be reevaluated. Tank failures have occurred because of pressure or vacuum resulting from inadequately sized or improperly maintained vents and overflows. The maximum withdrawal rate is usually assumed to be the value that occurs when the pipes at grade level break, or the maximum rate pumped from low-elevation reservoir tanks.

DEVICES FOR INDICATING WATER LEVEL

Some form of water-level indication should be provided on the tank so that operators can easily determine the water level. The most common devices used to measure water level are gauge boards and pressure transducer readouts.

Each form of water-level indication has advantages and disadvantages. Cost and the need for direct or remote reading, ease of maintenance, and performance in adverse weather conditions should all be considered when selecting an indicating system.

Gauge Boards

Gauge boards are normally composed of a float and target board on which direct water level indication is accomplished by noting the position of a target against a gauge board on the outside of the tank. The target marker is controlled by a cable attached to the float. As the float rises, the target marker is lowered. At capacity or full conditions, the target will be at the low end of the gauge board, making it potentially accessible to vandalism or snow buildup. Half-travel gauge boards are recommended in order to protect the target and float from vandalism and high winds. Float-type systems such as these are not recommended in freezing climates.

Remote Readings

A pressure transducer in the tank can indicate the water level at a remote readout some distance from the tank facility. The pressure transducer must be installed so that it is completely isolated from all inlet and outlet openings. Pressure transducers are sensitive enough to sense pressure changes created by water movement through a line that would cause a false reading. The pressure transducer can also control flow in and out of the tank by actuating pumps or valves.

Pressure Gauges

If freeze protection is provided, economical bourdon pressure gauges may be connected directly to the tank or riser.

Inlet Stop/Start Controls

A water utility may install a probe or transducer system to control the water level and to advise operators of low water levels. Probes require waterproof flanged entries at the top of the tank. In addition, for radio or wired telemetry equipment, an insulated conduit from the tank top to ground level must be installed to carry the electrical signal. If probes are used in tanks subject to icing conditions, the probe system should be designed to prevent damage from freezing.

Sampling Points

A sampling point near the tank bottom will allow the storage facility's water quality to be tested directly. Direct testing may be practical in low-usage tanks where the stored water may remain in the tank for a long time and water quality may deteriorate. Sampling points should be protected from freezing and vandalism.

EMERGENCY FILL/WITHDRAW CONNECTIONS

Some tanks may require provisions for emergency filling or emergency withdrawal. Typically, such provisions are needed when tanks are located in remote areas where fire protection groups may withdraw directly from the tank to fill pumper trucks or use stored water directly during firefighting. Where required, the emergency valving and connections should be designed to match the emergency facilities of the agency that will use them. The design must avoid cross-connections between the emergency system and the potable water system. Connections should be protected from freezing and vandalism, and the tank venting and overflow systems should be sized for these unusual fillings and withdrawals.

Chapter 3

Cathodic Protection

This chapter covers the theory, design, operation, and maintenance of cathodic protection systems. Corrosion of metals in contact with water is a natural process. Variables associated with a specific tank will determine how severe the corrosion activity will be. A properly designed, installed, and maintained cathodic protection system, used in conjunction with appropriate coatings, will provide economical and effective corrosion control for the interior submerged surfaces of a water-storage tank and the piping beneath the slab. For specific design details of cathodic protection systems, refer to ANSI/AWWA D106, Standard for Sacrificial Anode Cathodic Protection Systems for the Interior Submerged Surfaces of Steel Water Storage Tanks; ANSI/AWWA D104, Standard for Automatically Controlled, Impressed-Current Cathodic Protection for the Interior Submerged Surfaces of Steel Water Storage Tanks; NACE International Standard SP0388, Impressed Current Cathodic Protection of Internal Submerged Surfaces of Steel Water Storage Tanks; and NACE Standard SP0196, Galvanic Anode Cathodic Protection of Internal Submerged Surfaces of Steel Water Storage Tanks.

Cathodic protection is a method of corrosion control that can be used with all recommended American Water Works Association (AWWA) coating systems to prevent metal loss at any void in the coating below the water level. The design, costs, and potential benefits of using cathodic protection with coatings in an overall corrosion control program can be evaluated by qualified utility staff members, outside consultants, or cathodic protection equipment manufacturers.

NATURE OF CORROSION

The corrosion of steel in aqueous solutions is an electrochemical process in which a current flows and a chemical reaction occurs. A corrosion cell has four basic elements: anode, cathode, electrolyte, and closure path. The anode is the metal that will corrode; that is, metal ions leave its surface and enter the electrolyte solution. The cathode is a metal from which no metal ions enter the solution. The electrolyte may be any solution, such as drinking water, that is capable of conducting electricity. The closure path, also called the return current path, is the electrical conductor, usually metal, that connects the anode and cathode together. If any one of these elements

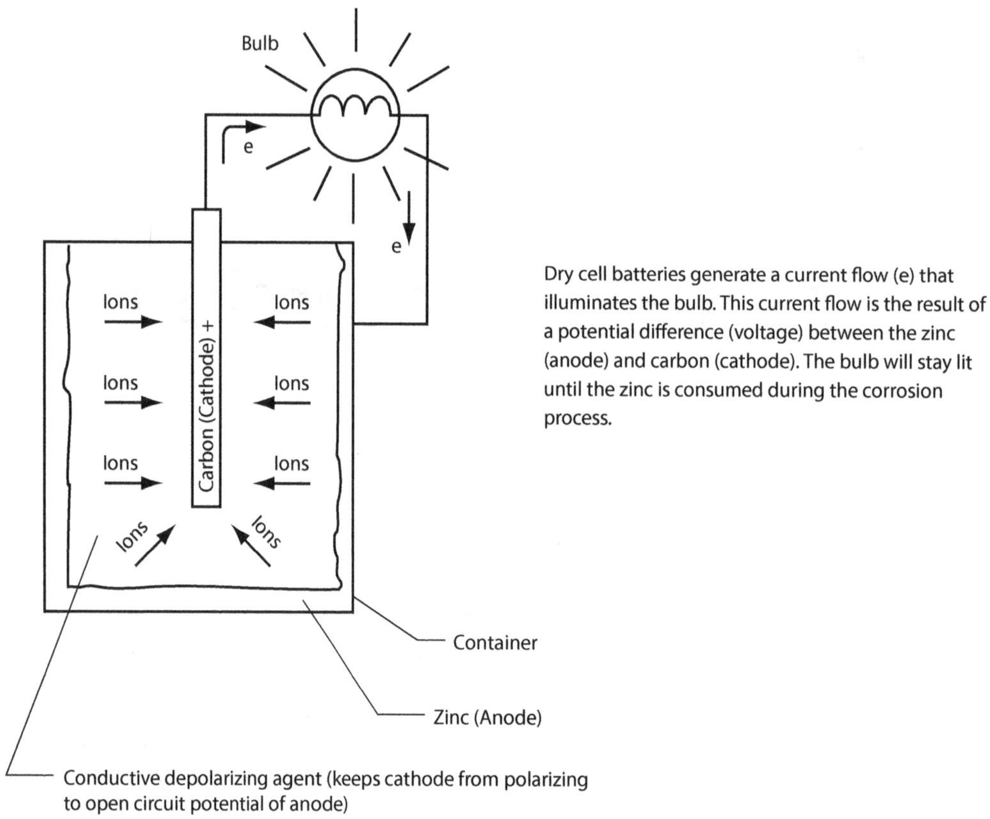

Figure 3-1 Schematic diagram of a battery

is missing, corrosion does not occur. For example, coating stops corrosion from occurring by providing a barrier to the current flowing between the metal and the electrolyte.

A dry-cell battery is a corrosion cell (Figure 3-1). When the battery's anode (zinc) and cathode (carbon) are connected through a closure path (the lightbulb), the potential difference between the zinc and carbon produces a current flow. The current will continue to flow until the zinc anode is consumed by the corrosion process.

It is important to consider why the current flows in the direction it does. The direction of flow is determined by the metals selected for the battery's case and center post. If the center post was magnesium instead of carbon, the current flow would be reversed. In this case, the magnesium center post would be the anode (which corrodes) and the zinc case would be the cathode (which does not corrode)

The current can also be forced to flow in the opposite direction if the standard carbon-zinc battery is connected to an outside current source (in place of the lightbulb). In this situation, the anode and cathode would also be reversed—that is, the battery case would become the cathode and would be protected from corrosion.

In a steel water-storage tank, some portion of the metal will be the anode and some portion will be the cathode. Which area takes on which function will be the result of such factors as impurities in the metal; surface conditions; oxygen concentrations in the water; the presence of any dissimilar metals; and stresses caused by manufactur-

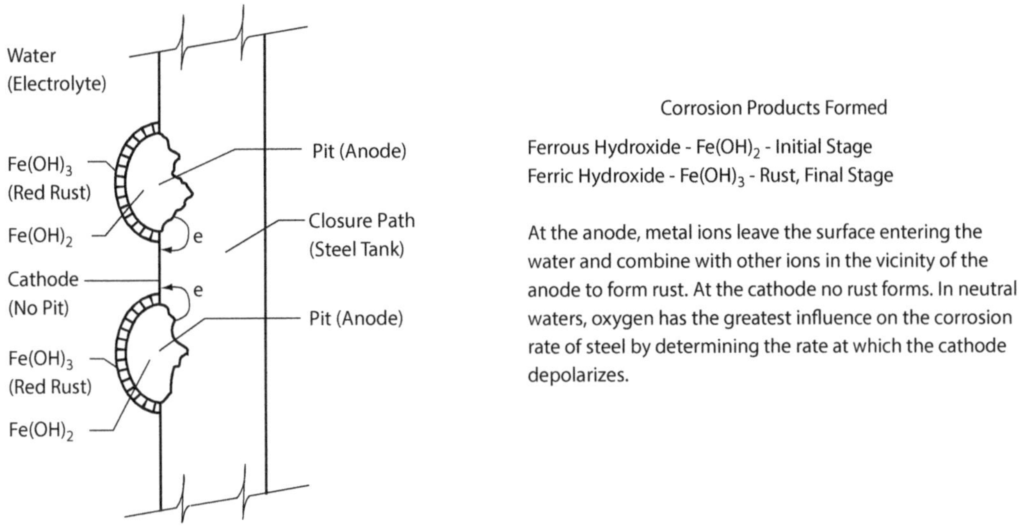

Figure 3-2 Corrosion of steel in water

ing, heat, or concentrated structural loads. Corrosion of steel in water is diagrammed in Figure 3-2. At the anode, metal ions leave the surface, enter the water, and combine with oxygen to form rust. Electrons released from the anode travel through the metal to the cathode. At the cathode, an ion exchange occurs but no metal is lost and no corrosion occurs.

The presence of ladders, floats, or other accessories made of stainless steel will cause steel exposed at holidays (voids) in the coating to corrode at a higher than normal rate. In such cases, the stainless-steel components are the cathodes, and the exposed steel is the anode.

For more detailed descriptions of corrosion chemistry, see AWWA Manual M27, *External Corrosion—Introduction to Chemistry and Control.*

PRINCIPLES OF CATHODIC PROTECTION

Cathodic protection systems are used to prevent or retard the corrosion that would naturally occur in a steel water tank. These systems prevent or slow corrosion by altering the electrochemical environment so that the submerged tank shell becomes the cathode of a corrosion cell. Since the cathode of a cell does not corrode, the submerged metallic tank shell is protected. There are two basic types of cathodic protection systems: galvanic and impressed-current systems.

Galvanic Systems

In a galvanic system, a block of specially selected metal, called a sacrificial anode, is immersed in the electrolyte and electrically connected to the metal of the tank. The metal of the sacrificial anode is selected so that it will become the anode of the corrosion cell, with the steel tank being the cathode. When blocks of magnesium are attached to steel structures and immersed in a conductive electrolyte (the water in the tank), the magnesium corrodes and the steel is protected. Galvanic protection systems have had limited application to water-storage tanks, although they may be used in climates that do not cause ice to form in the water stored in the tank. A description of the installation and performance was published in M. Lewis, "Cathodic Protection of 32 Steel Reservoir Interiors," *Materials Performance*, 35(9):13–17 (1996).

Impressed-Current Systems

In an impressed-current system of cathodic protection, an outside source of electrical power forces current into anodes submerged in the storage tank's water. The current flows from the anodes, through the water (electrolyte), and onto the submerged walls of the tank, making the tank itself the cathode of the corrosion cell. An impressed-current cathodic protection system consists of a manual or automatic alternating current/direct current (AC/DC) converter (i.e., a rectifier), feeder wires, and anodes located within the tank. Solar or wind-powered power sources may also be used. Output voltage is adjusted manually or automatically to control the DC output in order to account for a wide range of variables. Care must be exercised to ensure that the polarized voltage does not exceed a maximum value of approximately –1.05 V in reference to a copper-copper sulfate half cell; otherwise, the coating may be damaged. The precise maximum negative voltage is dependent on the characteristics of the coating and other factors.

Protective Coatings

Cathodic protection is normally used in conjunction with a well-coated tank surface. The coating reduces the rate of anode consumption and power use. Coatings usually have microscopic voids, which expose the metal to the water and allow metal loss if cathodic protection is not also in place. The ideal corrosion-control system combines a good dielectric coating (metallic coatings are not dielectric) and a properly designed, installed, and maintained cathodic protection system.

Exterior Corrosion

Cathodic protection systems are usually designed to protect the interior wetted surfaces of a water-storage tank. There are cases, however, where the exterior of a tank bottom or shell is in contact with corrosive soils. In those situations, proper selection of the tank base material or backfill may reduce corrosion, or a separate cathodic protection system can be designed to protect the tank exterior surfaces in contact with soil (refer to AWWA Manual M27, *External Corrosion—Introduction to Chemistry and Control,* for details). In cases where cathodic protection will be used to protect exterior tank surfaces in contact with the soil, protective coatings applied to the surfaces of the tank in contact with the soil will reduce the capital and operating costs of the cathodic protection system.

CATHODIC PROTECTION DESIGN

In designing a cathodic protection system, the engineer must consider the quality of the protective coating, tank geometry, surface area, obstructions, geographic location, temperature, turbulence, and chemical composition of the water stored in the tank. Some of the items to be specified in the design are the AC/DC converter or alternative power source, the anode materials, and the anode configuration and suspension.

Automatically Controlled AC/DC Converter (Rectifier)

The protective-current demands within a water-storage tank will continuously change because of variations in water chemistry, temperature, water-level fluctuations, coating deterioration, and polarization effects. Automatically controlled, impressed-current cathodic protection systems are typically used in water-storage tanks to adjust for these variations.

Reference electrodes are used to continuously monitor the protective level and control the amount of cathodic protection current delivered to the structure by the system. Separate control circuits are used for riser pipes and other areas of the tank that may have different localized conditions.

Anode Materials and Design Life

Impressed-current anodes may be made of aluminum, which usually has a design life of 1 to 2 years; or they may be made of high-silicon cast iron, platinum, mixed metal oxide, or other long-life anode materials that can provide a nominal 10- to 20-year life. The design life of the anode system is based on the anticipated protective-current requirements and the known consumption rates for the selected anode material. The approximate consumption rates for selected anode materials are

Aluminum	14 lb/year (6.4 kg/year)
High-silicon cast iron	1.0 lb/year (0.45 kg/year)
Platinized niobium, platinized titanium, or mixed metal oxides on titanium	0.00008 to 0.0013 lb/year (0.036 to 0.59 g/year)

Anode Configuration and Suspension

Current distribution from any anode configuration is affected by the geometric shape of the tank, obstructions within the tank, interior coating, and chemical characteristics of the water.

The anode system may be installed vertically or horizontally. Vertically suspended anodes are installed by hanging the anode from an electrically insulated device at the tank roof, adjacent to holes cut in the roof (Figure 3-3). Horizontally suspended anodes are positioned below the normal water level, attached to the tank shell or access tube (Figure 3-4). In elevated tanks with an inlet-outlet riser pipe of 30-in. (76-cm) diameter or larger, a vertically suspended anode is used to provide protection within the riser.

When holes are cut in the roof, the finished installation must be watertight to eliminate openings for insects and runoff to enter the tank.

For tanks subject to icing, either vertical anode systems with extensible elements or horizontal suspension systems designed to minimize ice damage to the anodes should be considered. These suspension systems can provide year-round protection and may eliminate the need for annual anode replacement due to ice damage.

MAINTENANCE

Following the installation of a cathodic protection system, a post-installation tank-to-water potential profile is performed to ensure that the system is providing optimal corrosion control. The level of corrosion control achieved by the cathodic protection system can be determined through electrical testing. Corrosion is under control when a copper-copper sulfate reference electrode is placed adjacent to, but not touching, the submerged tank surface and a polarized tank-to-water potential of –0.85 V or more negative is measured.

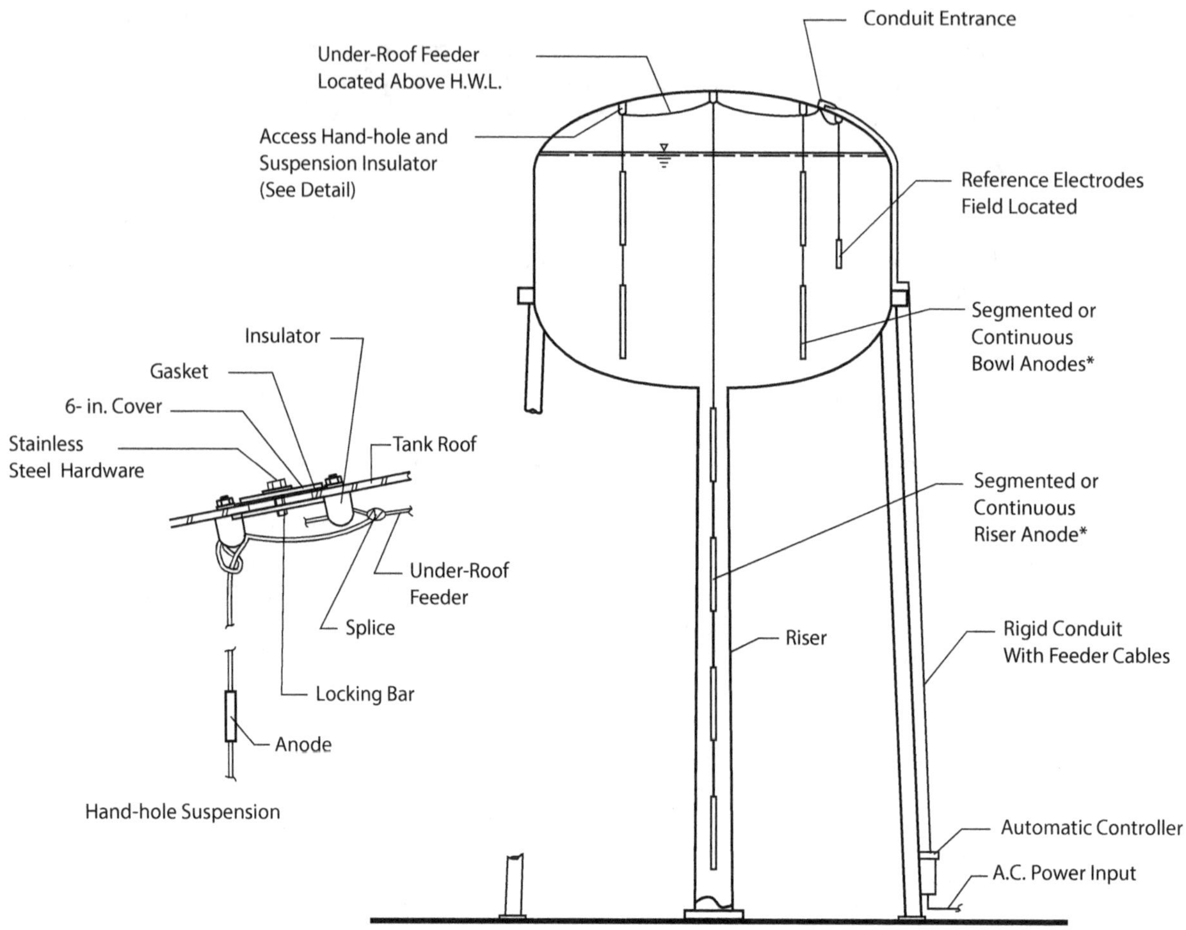

*Type, quantity, length, and location of the anodes and reference electrodes are determined by engineering design to suit individual applications.

NOTE: Typical anode materials used for icing and nonicing climates: platinum niobium, high-silicon cast iron.

Figure 3-3 Tank corrosion protection—vertically suspended anodes

Manually controlled rectifiers require periodic adjustment to maintain the system within an optimal protective range. The manufacturer's instructions describe how to adjust rectifiers. The cathodic protection designer will establish the range of operation most suitable for each application based on a tank-to-water potential survey.

Automatic rectifiers will continuously monitor the tank-to-water potential being maintained by the system and make adjustments to control corrosion. Personnel responsible for operating and maintaining the cathodic protection system should refer to the designer's instructions to fully understand their responsibilities. They should consult with the manufacturer if necessary regarding the equipment's operation and make certain that all responsible personnel are familiar with its operation. A successful cathodic protection corrosion-control system will continuously operate within the established criteria.

Annual inspection of the cathodic protection system by the manufacturer or by a qualified corrosion engineer is recommended. At a minimum, this inspection should

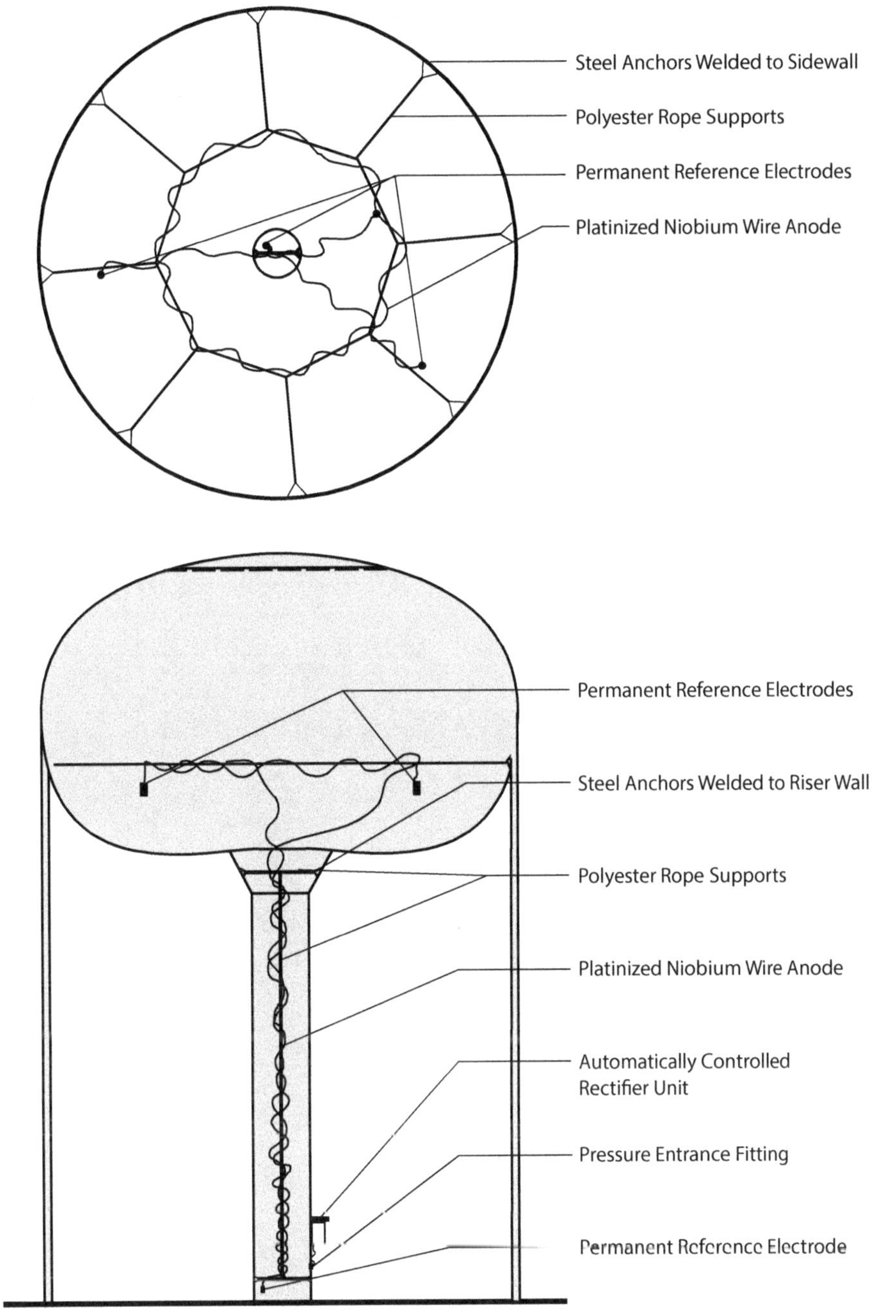

Figure 3-4 Tank corrosion protection—horizontally suspended anodes

include an overall examination of the entire cathodic protection system, replacement of all defective parts, a potential profile survey, a physical check of the anode placement, and a written report. For tanks in cold climates, the annual inspection should be performed in the spring.

Various annual service plans are available from the cathodic protection companies or other service organizations. Cathodic protection systems should be tested and inspected on a regular basis to ensure that they provide the maximum level of corrosion control to the submerged steel tank surfaces.

This page intentionally blank.

Chapter 4

Coating Systems

When exposed to the environment, steel oxidizes and deteriorates. This is especially true when the environment includes both oxygen and moisture. For steel water tanks, paints and other protective coatings are used to prevent such deterioration. This chapter covers coating systems used to protect the interior and exterior of steel water-storage tanks.

A protective coating is a material that, when applied to a structure, will isolate the structure from its environment. Properly applied protective coatings are a cost effective way to protect both exterior and interior tank surfaces. A coating applied to the interior surfaces of a tank is also called a lining.

Both exterior and interior coating systems must be carefully selected to provide the best protective value for the money based on coating life, effectiveness of protection, ease of application, and ease of adding coats to the system in future years. Protective coating systems have become much more complex than the single component materials that were prevalent prior to 1970.

Safety precautions and special equipment such as safety harnesses, safety lines, protective clothing, and air-supplied face masks are required to protect workers. Poor visibility and substrate access difficulties make abrasive blast cleaning and coating application on tank interiors extremely difficult. The storage structure's design may also include inaccessible areas, stitch welds, lapped roof sections, or other configurations that make it more difficult to obtain a proper coating application. A coating system that requires ideal conditions to achieve proper application may not be appropriate, no matter how effective the final coating would be if properly applied.

INTERIOR COATINGS

Protective coating systems intended for use on the interior surfaces of potable water storage tanks must be able to withstand the following:

- constant immersion in water
- varying water temperatures
- alternate wetting and drying periods
- ice abrasion in cold climate areas

- high humidity and heat in the zones above the high water level
- chlorine and mineral content of the water

An additional requirement for such materials to be applied successfully in the field is application latitude—that is, the suitability of the material to be applied under extremely adverse conditions.

Approval by Regulatory Agencies

Coating and lining materials intended for use in contact with potable water must not impose a health risk on the general public and must be approved for such use by the appropriate state or federal regulatory agency. Although some state health departments evaluate and approve tank-lining materials themselves, the majority of state health departments have relied on specific product approval by NSF International since the American Water Works Association (AWWA) does not test or approve individual products.

Because most states had placed the burden of approval on the federal government, the backlog of materials waiting for approval had increased substantially. In response to this problem, the US Environmental Protection Agency (USEPA) solicited proposals from nonprofit organizations to implement a testing and evaluation program for materials in contact with potable water. A consortium of organizations led by NSF International reached an agreement with USEPA and developed this program, resulting in the publication of NSF/ANSI Standard 61, Drinking Water Systems Components—Health Effects. Other approved testing laboratories may test and certify products for compliance with NSF/ANSI 61. Contact ANSI for a list of approved labs.

Field-Applied Interior Coating Systems for Welded Steel Tanks

ANSI/AWWA D102 recognizes general types of interior coating systems: epoxies and polyureas. Each of these systems has provided satisfactory service when correctly applied. As a result of differing climates, governmental regulations, and other variables, no one type of coating system can be considered best. Epoxy systems are currently the most popular. Coatings such as polyureas and plural-component urethane materials are also being used.

Epoxy coatings. A variety of epoxy coatings is available. The epoxies used by municipal water utilities are usually two-component polyamide cured products.

Epoxy coatings are popular because of their outstanding impermeability to water migration through the membrane, resistance to ionic transfer, outstanding adhesion, high film build, and resistance to cathodic disbondment. Additionally, the high solids and low volatile organic compound (VOC) content of many of the epoxy formulations may comply with current government regulations regarding both air pollution during application and VOCs in potable water during operation.

Two-component epoxies have several limitations. Cold temperatures may retard or stop their curing action if special cold-weather formulations are not used. The two components must be mixed correctly. And, finally, most thoroughly cured epoxy coatings develop very hard films that require scarification prior to recoating.

Polyurea coatings. Due to recent advancements in the development of 100 percent solid coatings, polyurea coatings are now commonly used for the lining of water storage tanks.

Polyurea coatings tend to be more expensive than the traditional two-component polyamide epoxies due to higher material costs and higher application costs, including specialty spray pump equipment. However, these higher costs are offset by the coating's ability to cure quickly, even at low temperatures and high humidity. Polyurea coatings also have excellent tensile strength and abrasion resistance. Additionally, since the polyurea coatings are typically 100 percent solids they will comply with VOC regulations.

Factory-Applied Interior Coating Systems for Bolted Steel Tanks

The interior surfaces of factory-coated bolted tanks may be coated with one of the following: thermoset liquid epoxy, thermoset powder coating, galvanizing, or fused glass. Factory coating systems are uniformly applied and fully cured under controlled conditions. The thermoset liquid epoxy coatings are two-component systems applied in two coats, then baked on at approximately 425°F–525°F (218°C–274°C). The thermoset powder coatings are applied in one coat and baked in accordance with the manufacturer's instructions. Galvanized coatings are applied in accordance with the American Hot Dip Galvanizers Association in compliance with ASTM A123 and ASTM A153. Fused silica glass coatings are achieved by spraying a ground-glass-and-water slurry onto the steel panels, then firing the panels in a furnace at approximately 1,600°F (870°C). Coating systems for factory-coated bolted steel tanks are discussed in more detail in ANSI/AWWA D103.

EXTERIOR COATINGS

Factors to consider in selecting a coating system for the exterior surfaces of a tank include

- the type of atmosphere in which the tank will be located
- the area surrounding the tank
- the expected ambient temperatures and prevailing winds during the time of year when the coating project is scheduled to be performed
- the appearance of the coating

Appearance and aesthetic value are particularly difficult to evaluate. Nonetheless, the tank's appearance is of great importance to people living or working near it. A tank is also a widely visible statement of a municipality's prosperity, cleanliness, and spirit.

Field-Applied Exterior Coating Systems for Welded Steel Tanks

The coating specifier should consider the characteristics of the various exterior coating systems. For example, elevated water-storage tanks in congested areas need coating materials that can be roller applied or will dry in the time it takes overspray particles to fall from the tank to the ground. Another special case is coastal environments that require coating materials that are more resistant to the salt-laden atmosphere and provide a higher film build than is provided by some of the conventional coating systems.

Coating systems and colors can be combined on different portions of a tank to achieve good corrosion protection while creating an attractive tank.

Coating is not recommended for the undersides of welded steel reservoir and standpipe tank bottoms. Coating applied prior to or during installation will not be continuous and will therefore promote corrosion at the bare areas. A totally bare tank bottom in uniform contact with the supporting material will be more corrosion-resistant than one that is coated but has voids in the coating. Refer to the section on exterior corrosion in chapter 3 for information concerning cathodic protection and coatings on the soil contact surfaces.

Factory-Applied Exterior Coating Systems

The exterior surfaces of factory-coated bolted tanks may be coated with one of the following: galvanizing, fused glass, thermoset liquid coatings, or thermoset powder coatings. In all cases, the steel panels are coated following roll forming and bolt-hole punching. Galvanized coatings are applied with zinc metal in compliance with ASTM A123 and ASTM A153 after the parts have been fabricated. Silica glass coatings are fused to the steel tank panels by furnace firing at 1,600°F (870°C). Liquid baked-on coating systems combine an epoxy primer with either an acrylic enamel or acrylic urethane topcoat. Baked-on powder coatings are epoxy, acrylic, or urethane composition. The exterior colors should be selected when the tank is ordered.

INSPECTION AND QUALITY CONTROL

Although defective coating material can cause the protective coating system to fail, it is rarely the reason a protective coating system fails. More common reasons for failure include

- incorrect system chosen for a service
- inadequate surface preparation
- improper application procedures
- inadequate cure
- inadequate ventilation during application, drying, and curing
- improper mixing
- unacceptable ambient conditions

A complete list of causes for coating failures is quite extensive and includes a number of items that would not be intuitively suspected. For example, the depth of the surface profile (roughness of substrate) can affect the amount of shear force a coating can withstand, which will then affect the integrity of the coating system.

Because special knowledge is needed to evaluate the job and failed coatings can be costly, a qualified third-party firm should inspect the coating application on the owner's behalf. The firm selected should be trained and experienced in tank coating. Firm personnel must also be willing and able to climb and use rigging equipment to reach the work area. Owners can hire qualified independent inspection firms to perform periodic maintenance inspections of existing tanks and for quality assurance inspections of any work in progress. Competing coating contractors or coating suppliers may have a conflict of interest and are not recommended as inspectors.

REMOVING COATING BY ABRASIVE BLASTING

The removal of regulated heavy-metal coatings, such as chromium or lead-based paints, from water-storage tanks has long been a concern. The use of silica sand as an abrasive has been restricted in some areas of the country due to the health risks caused by the release of free silica into the atmosphere. In addition, many of the coatings on water storage tanks contain lead. State and local USEPA regulations restrict the levels of lead and silica to which the public can be exposed. New blast abrasives and creative methods of containing abrasive blast residue have been necessary to comply with these restrictions.

Most states have "nuisance" regulations, and several states have implemented tighter restrictions on the abrasive blasting of steel structures such as water tanks. These restrictions include the Clean Air Act, which currently does not allow more than 1.5 $\mu g/m^3$ of lead per day to be released into the atmosphere over a 90-day period. In addition, the National Ambient Air Quality Standard currently does not allow more than 150 $\mu g/m^3$ of particulate matter that is less than 10 μm in size to be released into the atmosphere over an 8-hour work day. Enforcement of these regulations will not allow open blasting to be performed as has been done in the past.

Although some governing entities may rely only on their "nuisance" regulations, the owners, specifiers, and contractors are still responsible for protecting the environment and personal property. It should be noted that lead or other hazardous materials do not have to be present in the coating system for compliance with present regulations to be required. Many states are now enforcing "nuisance" regulations, the Clean Air Act, or other regulations that, in the past, had not been considered or made any requirements for compliance. It is still necessary to control nuisance dust as well as particulate matter. The presence of lead or other hazardous materials increases the potential for contaminating public or private property and adds to the liability of each recoating project. Therefore, sound engineering judgment must be used to determine the best method to remove the existing coating, and each project must be researched to determine the regulations that will affect the project.

Waste Disposal

At the conclusion of an abrasive blasting project, the blasting debris must be tested to determine whether or not it is hazardous. A specified number of random samples must be taken and tested in accordance with the USEPA's Toxicity Characteristics Leaching Procedure (TCLP) test. The sampling procedure requires mapping the debris containers and using a *random* method of determining the locations where the samples will be taken. USEPA has requirements for sampling techniques, and these techniques are recommended.

After the TCLP tests have been performed, the leaching characteristics of the blast debris can be determined. Often the debris from a lead paint removal project has to be disposed of as potentially hazardous waste. The implementation of USEPA's Land Ban restrictions prohibits the disposal of much of the abrasive blast residue in hazardous waste dump sites. The Land Ban requires that debris that is tested and found to contain more than the allowable levels of leachable lead must, in many cases, be stabilized or have the lead extracted prior to disposing of the materials. These added restrictions on disposal of the abrasive blast residue further escalate the already high costs of removing lead-based coatings from water-storage tanks (see Resource Conservation and Recovery Acts [RCRA] regulations 40 CFR 240 through 280).

For the handling and disposing of materials, many states or political governing agencies have imposed restrictions that are more stringent than federal regulations require. Additionally, the owner, engineer, and constructor must be aware of transportation restrictions affecting the hauling of hazardous wastes to treatment facilities and waste sites. All manifests and other documentation must be in order to keep a "paper trail" of the waste material, because the owner is responsible for the debris forever. The owner will significantly reduce potential liability by properly documenting every step.

Several methods are available to minimize the amount of hazardous or potentially hazardous waste materials to be disposed. These methods include the use of recyclable abrasives, water jetting, and abrasive additives; many more technologies are being developed.

Part II

The New Tank Project

This page intentionally blank.

Chapter 5

Selecting and Sizing Water-Storage Tanks

The selection and sizing of a water-storage tank involve a number of engineering considerations and generally require a detailed analysis of water demands, supply sources, and the distribution system. The purpose of this chapter is to discuss these design parameters and factors to consider in selecting and sizing a steel tank. A detailed treatment of each factor has not been attempted.

PEAK DEMAND

Peak demand is usually the first factor to consider when sizing a distribution system tank. Most water supply sources are best operated on a 24-hour production basis and produce a quantity of water in 24 hours that is equal to the 24-hour demand. Although clearwells offer a cushion between production and demand, clearwell capacity is usually considered production reserve rather than distribution reserve. If distribution system supply sources are operated with a relatively constant pumping rate equal to the daily demand rate, any water in excess of the hourly demand must be stored in elevated tanks (whether the elevation is natural or structural). The usual curves for demand are lowest in the early morning hours, and the tanks are filled during this period. As the day progresses, demand increases and usually peaks in late afternoon; the tanks feed back into the system during this period. A tank functioning in this manner helps maintain a relatively constant pressure in the system.

Figure 5-1 shows a typical daily demand curve. In this example, the maximum consumption rate is 200 percent of the average daily rate, and the quantity stored to achieve a level pumping rate is 20 percent of the daily consumption. This 20 percent of daily consumption is not necessarily the optimal ratio of storage to consumption, because most water regulatory agencies require more storage or emergency sources.

FIRE FLOW

Fire flow is usually the second factor to consider when determining tank capacity. Insurance underwriters have developed formulas to determine desirable quantities, pressures, and flow duration. Using these formulas, all classes and uses of all buildings within the area served are considered. Frequently, storage requirements for fire flow are greater than the storage required for system regulation, and a large fire-flow demand may require additional pumping capacity as well as the use of stored water.

TOP AND BOTTOM CAPACITY LEVELS

In addition to establishing the storage facility's necessary capacity, required top and bottom capacity levels must also be established. These three values, combined with aesthetic and economic considerations, greatly influence the geometry of the final tank design.

A detailed hydraulic analysis of the water distribution system for which the storage tank is being designed is usually conducted to establish the BCL and TCL elevations that will provide effective, functional storage at a given tank site. A distribution system is analyzed by creating a computer model with hydraulic data for the pipeline distribution network, distribution system pumping facilities, and various water demand conditions. This program produces hydraulic gradients across the distribution system for the particular demand condition represented (e.g., fire flow, maximum hour, maximum day, tank replenishment). The goals of this exercise are (1) to produce a coordinated design covering system pumping capacity, head conditions, and pipeline improvements necessary to provide adequate system transmission capacity to and from the tank, and (2) to establish the range of operating gradients or water levels at the tank. For further information, refer to AWWA Manual M31, *Distribution System Requirements for Fire Protection*, and AWWA Manual M32, *Computer Modeling of Water Distribution Systems.*

WATER QUALITY ISSUES

The design phase of a new tank project is the best time to consider how a tank design and piping configuration may contribute to water quality. Water circulation and water flow should be included in the design parameters. Water age can be managed through a well-designed system that "exercises" the tank; considerations include water turnover, altitude valves, pumping management, and other components for maintaining fresh water in the tank and mitigating water quality issues. Both active and passive systems for improving water quality in tanks are available for new and existing tanks.

Water quality is a significant concern to water distribution system managers. Disinfection by-product (DBP) formation is largely dependent on reaction time, and it can continue for several days within the distribution system. At the same time, disinfectant residual must be maintained throughout the most remote components of the system to ensure pathogen-free water. Managing the residence time of water within storage tanks is one practice available to minimize water age within the distribution system. Water system managers and engineers should consider the need for circulation of water and residence time management within storage tanks during the design phase.

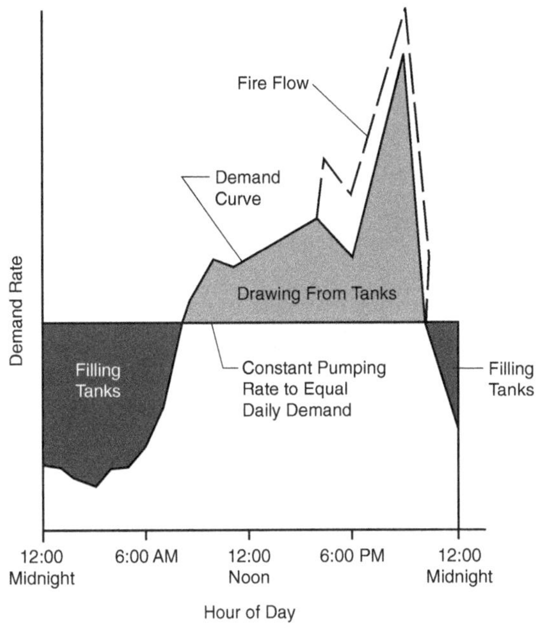

Figure 5-1 Typical daily flow at constant pumping rate

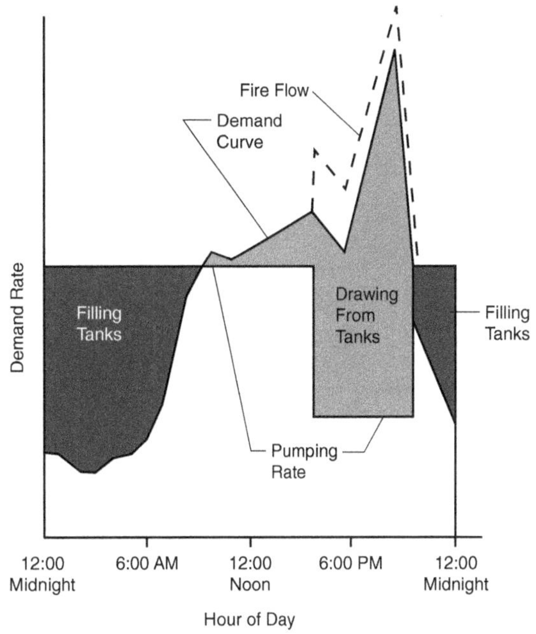

Figure 5-2 Typical daily flow with variable-rate pumping

ENERGY COSTS

Many power utilities have adopted rates based on when electricity is used, and it may be cost-effective to control pumping in an effort to reduce the maximum power demand. Figure 5-1 shows the use of constant-rate pumping for 24 hours. However, the part-time use of more or larger pumps may be more cost-effective. To get the best electric power rates, pumping (or power-demand load) must be reduced during the periods when the maximum electrical demand occurs. A simple way to reduce rates is to pump more water into the storage tanks during hours when electrical power demand is low and to reduce pumping during periods when that demand is high (Figure 5-2). This technique requires increased storage capacity.

FUTURE NEEDS

Future needs are an important consideration, and where practical, a tank should be sized to provide for anticipated future growth and the resulting increase in water demands. This consideration is particularly important in the design of water-storage tanks, since they represent a large capital investment, and future enlargement of their storage capacity is not always feasible. Proper sizing of a storage tank must also establish proper water turnover and circulation to ensure that water quality standards are met.

ENVIRONMENTAL IMPACT

The major environmental impact of the tank itself is its appearance. This impact can be mitigated by the use of tank designs and exterior coating systems that blend into the surrounding terrain. Site location and site development (discussed in detail in chapter 6) are also important factors to consider in reducing any adverse environmental impacts.

The increased availability of water for use by customers made possible by greater storage may also be an environmental concern, but this consideration relates to the broader topics of land-use planning and wastewater discharge capacity, which should be evaluated before the need for additional storage is addressed.

TANK COSTS

Tank costs vary with type, capacity, and site. These factors are interrelated and are discussed in the following paragraphs.

Variation With Type

The prime influence on cost is the configuration of the tank—i.e., whether a standpipe, reservoir, or elevated tank. Figure 5-3 illustrates the relative differences in the cost per unit volume for the three tank configurations. It is apparent that if an accessible high-elevation site is available, a reservoir-type tank will be the most economical.

The cost of a standpipe depends on its ratio of height to diameter. A tall, small-diameter standpipe will cost more than one of the same capacity having a diameter only slightly greater than its height. Two elements influence this cost differential. First, the minimum weight of steel to contain a given capacity is usually found in tanks that have a diameter equal to their height. Second, taller tanks cost more per unit weight of steel to erect because of the difficulties in lifting the steel and conducting assembly operations at greater heights.

When the cost per unit volume of a standpipe is computed, only part of the total storage may be considered effective storage. The designer should determine the head range within which the water is useful and compute from this the amount of effective water storage. The comparison of standpipe costs should then be based on cost per unit volume of effective storage.

Variation With Capacity

With elevated tanks, the cost per unit volume decreases significantly as the tank capacity is increased. A 100,000-gal (380,000-L) elevated tank has approximately twice the cost per unit volume of a 500,000-gal (1.9-ML) elevated tank (Figure 5-4). For reservoirs and standpipes, an increase in capacity also lowers the cost per unit volume, but the unit cost levels out at a capacity of approximately 5 mil gal (19 ML).

Variation With Site

The importance of a well-conceived site location cannot be overemphasized. Access costs, construction costs, foundation costs, and insurance costs can all be minimized if the site selection guidelines set forth in chapter 6 are followed.

Cost Estimates

As improvements are made in methods of design and construction, and as competitive market forces change, the pricing guidelines will change. This will affect the accuracy of Figures 5-3 and 5-4. Current estimates of construction costs should be obtained from tank contractors before a tank size, configuration, or style is selected.

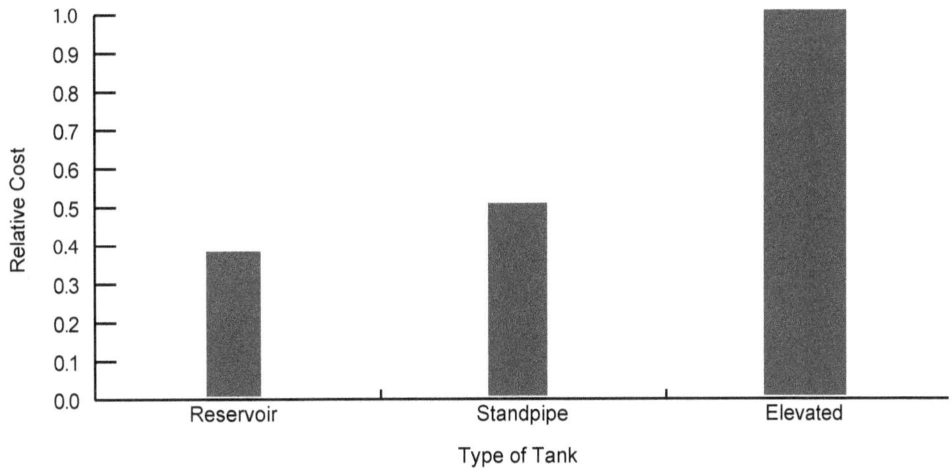

Figure 5-3 Relative cost by type of steel tank for 500,000-gal (1.9-ML) tanks

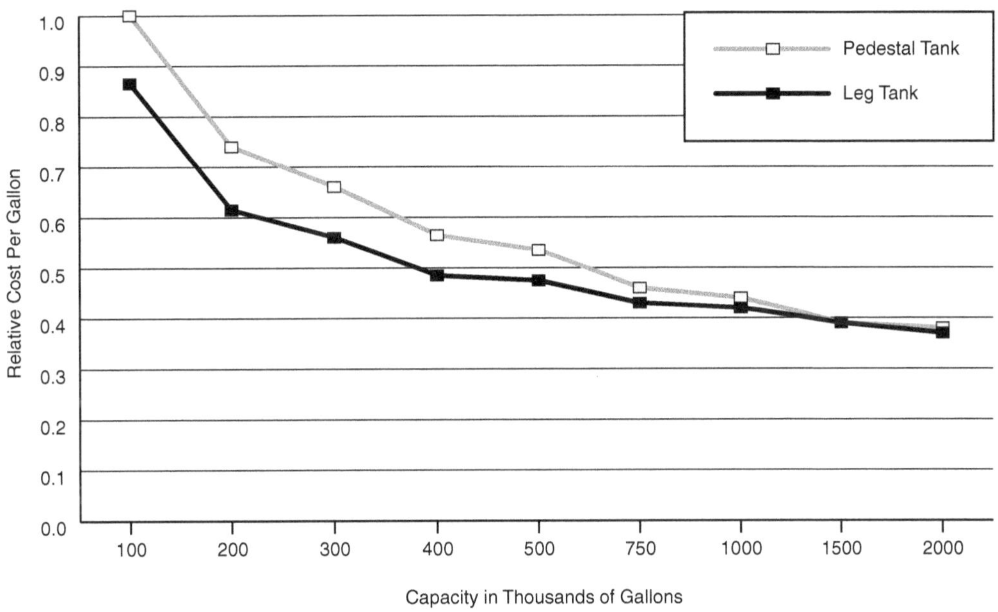

NOTE: Horizontal axis not to scale.

Figure 5-4 Relative cost by type of elevated steel tank

This page intentionally blank.

Chapter 6

Construction Considerations

This chapter includes general discussions of each of the topics to be investigated and detailed in the specifications for a new steel water-storage tank. It also describes the content and purpose of each of the documents involved in bidding a tank, and it notes the duties of the owner, the engineer, and the constructor with respect to specifications, bidding, and design.

DESIGN STANDARDS

ANSI/AWWA D100, Standard for Welded Carbon Steel Tanks for Water Storage, ANSI/AWWA D103, Standard for Factory-Coated Bolted Carbon Steel Tanks for Water Storage, and ANSI/AWWA D107, Standard for Composite Elevated Tanks for Water Storage, assign specific responsibilities to the owner (or the owner's engineer) and the constructor for the construction of steel tanks. The requirements for the design, water, snow, wind, and live loads in all of these tank standards are the minimum allowable values. The engineer is expected to be aware of all local codes and ordinances and should provide the required loading information, design information, or both to the constructor if local requirements are more stringent than those found in the applicable AWWA standards.

Standards for seismic design are constantly changing. Therefore, the specifier must refer to the latest editions of AWWA standards and local building codes to properly specify the appropriate seismic design criteria for the tank to be purchased. Different approaches exist for determining allowable loads for the buckling of conical and dished sections, and the approaches used may be limited by the owner or engineer. New standards and clarifications to the design approaches are being used. Design methods must consider construction tolerances.

CONTRACT DOCUMENTS

The contract documents should include sufficient data to allow all qualified constructors an equal opportunity to bid for the tank. The completed soil survey should be accomplished by a qualified geotechnical engineer so that all bidders are using identical data. After the contract is awarded, the manufacturer or constructor should prepare foundation plans and detailed steel tank drawings, which should be submitted to the owner for review before any fabrication begins.

CONSTRUCTOR CAPABILITIES

Tank constructors specialize in the design, fabrication, and erection of welded or bolted steel storage tanks; they must rely on specialty constructors to provide specialized equipment. A local supplier may be able to provide appurtenances (for example, electrical level-control systems); however, a tank constructor may select a familiar supplier located a considerable distance from the project. Dividing portions of the main contract into specialty work functions encourages the interest and participation of local suppliers. Using local suppliers provides employment in the area and reduces costs by minimizing the required travel. Any portions of the new tank construction project that require cutting and welding on the tank should be done by the tank constructor's own forces or with the tank constructor's written approval.

Often a piece of equipment that needs to be installed on the tank must interface with another piece of operating equipment in a water system (for example, telemetering systems). One constructor should provide the entire electrical or mechanical system in order to minimize coordination between constructors and interface problems with different equipment.

GUARANTEES

A guarantee for a specific time after the work is completed should be given on the structure, coatings, appurtenances, and other work, as specified by the purchaser. Normally a one-year period in which all workmanship and materials are warranted is included in the owner's specifications.

SOIL INVESTIGATIONS

The soil investigation on which the foundation design is based is critical to the long-term stability of the foundation. Soil loadings vary with the type of tank. A 100-ft (30.5-m) high standpipe, for example, exerts a contact pressure of 6,240 lb/ft^2 (229 kPa), whereas a 1-mil gal (4-mL) capacity elevated tank and its contents produce a total load of approximately 10 million lb (45 MN) that must be transmitted to the soil. The forewords of ANSI/AWWA D100, ANSI/AWWA D103, and ANSI/AWWA D107 suggest that the owner have a soil investigation conducted to provide the data necessary to design the tank foundation. The soil investigation should be completed prior to the taking of bids, so that all constructors will have the same design criteria available and better competition will prevail. Table 6-1 lists the typical requirements for soil investigations, and Figure 6-1 shows a soil-testing operation. Unusual site conditions may dictate the need for additional soil investigation.

Table 6-1 Typical soil investigation requirements

Tank Type	Storage Capacity *gal (ML)*	No. of Borings	Boring Depth* *ft (m)*
Leg-Type Tanks	30,000–150,000 (0.114–0.568)	3	1 at 25–35 (7.6–10.7) 2 at 20–25 (6.1–7.6)
	200,000–500,000 (0.757–1.893)	4	1 at center, 25–35 (7.6–10.7) 3 at perimeter, 20 (6.1)
	>500,000 (>1.893)	5	1 at center, 25–35 (7.6–10.7) 4 at perimeter, 20 (6.1)
Single-Pedestal Tanks and Standpipes	≥50,000 (≥0.189)	2	1 at 25–35 (7.6–10.7) 1 at 15–20 (4.5–6.1)
	>50,000 (>0.189)	3	1 at 30–40 (9.1–12.2) 2 at 20–30 (6.1–9.1)
	≥500,000 (>1.893)	4	1 at 40–45 (12.2–13.7) 3 at 25–35 (7.6–10.7)
Ground Storage Tanks	Diameter ≥40 ft (12.2 m)	2	Depth equal to tank height; 20-ft (6.1-m) minimum
	Diameter >40 ft (12.2 m)	3	Depth equal to outside diameter of the footing or a minimum depth with an additional boring required for each 30-ft (9.1-m) increment in diameter greater than 40 ft (12.2 m)

*If preliminary field information indicates that the allowable net bearing pressure is less than 1,500 lb/ft^2 (720 kPa), at least one boring should go to sufficient depth to obtain piling information or two borings to sufficient depth to obtain caisson information.

A comprehensive soil investigation report will include the following items:

1. Information on the presence of rock, old excavation, or fill.
2. A detailed boring log complete with blow count records.
3. Record of any undisturbed samples.
4. Laboratory analysis of samples as required.
5. The classification of soil strata after appropriate sampling.
6. The homogeneity and compressibility of the soils across the tank site.
7. An estimate of uniform and differential settlements.
8. Water level in boring holes at time of drilling and before site is left.

9. The elevation of groundwater and whether dewatering is required.
10. The elevations of the existing grade and other topographical features that may affect the foundation design or construction.
11. The elevation of borings and exact location of borings on the site.
12. Whether the site is suitable for the structure to be built thereon, and what remediation, if any, is necessary to make it suitable.
13. A recommendation on the type(s) of foundation system(s) that will be required at the site.
14. If a deep foundation will be required, information on the type, capacity, and required length of piles, caissons, or piers.
15. Information as to whether allowable bearing values can be increased for wind, earthquake, or other temporary loadings; factor of safety for bearing values stated.
16. The bearing capacity of the soil, factor of safety used in obtaining bearing capacity, and depth at which the foundation must be founded.
17. For standpipes and reserviors, the minimum allowable foundation width for continuous and isolated footings, is applicable.
18. Information required for pseudodynamic seismic design of structure.

Figure 6-1 Soil-testing operations

The purchaser's soils report should be used as the basis for preparing bids and construction design. Any different soil conditions discovered during construction should be handled by change order, with the owner receiving credit for savings or bearing the cost of additional construction expenses. The constructor can commission an additional soil investigation at his or her own expense if he or she believes it is warranted. If the additional borings reveal conditions differing from the purchaser's report, then costs incurred by the necessary foundation design changes should be handled by change order.

Investigation Recommendations

A log of borings does not provide sufficient data for a foundation design. The purchaser should employ a qualified geotechnical engineer to evaluate the subsurface conditions for a particular site, relying on the soils consultant's familiarity with the area. After field exploration and laboratory testing, the soils consultant can recommend a foundation system for the structure. The soils consultant should determine what foundation systems are feasible for the site's soil and then select one based on the availability of foundation materials and constructors with the necessary equipment and expertise. Soils consultants will find that tank constructors can often assist in determining the type of foundation.

Proprietary types of foundation elements and soil improvement systems are available; however, the smaller size of many tank projects does not justify the costs of the specialty constructors who could implement such systems. The best technical solution may be inappropriate from a cost standpoint.

Responsibility for a Foundation Design

The project soil investigation, AWWA standards, and the technical sections of the project specification serve as a basis for foundation design by the constructor. With this information, the constructor can design the most economical foundation for the tank in compliance with recognized design methods and building codes. The tank supplier or constructor is ultimately responsible for the design of the foundation. Although some constructors prefer that the owner provide the foundation, others prefer to control and administer all activities that significantly influence the success of their project. These firms see the need to control construction of the foundation, since adverse foundation behavior could precipitate structural failure of the tank.

When a foundation has been designed by others for a particular project, the constructor must be allowed to review the design and make changes deemed necessary to ensure that the foundation will adequately support the tank both dimensionally and structurally. Variations will necessitate revisions of purchaser's drawings and possibly cause confusion and errors in administering and constructing the foundation.

RESERVOIR AND STANDPIPE FOUNDATIONS

ANSI/AWWA D100 allows five alternative types of foundations for flat-bottomed welded steel tanks: type 1, tanks supported on ringwalls; type 2, tanks supported on concrete slabs; type 3, tanks within ringwalls; type 4, tanks supported on granular berms; and type 5, tanks supported on granular berms with steel retainer rings. ANSI/AWWA D103 adds a sixth alternative for bolted steel tanks: type 6, tanks with sidewall embedded in concrete slab. This last type uses a reinforced concrete slab for the tank bottom in lieu of steel plate.

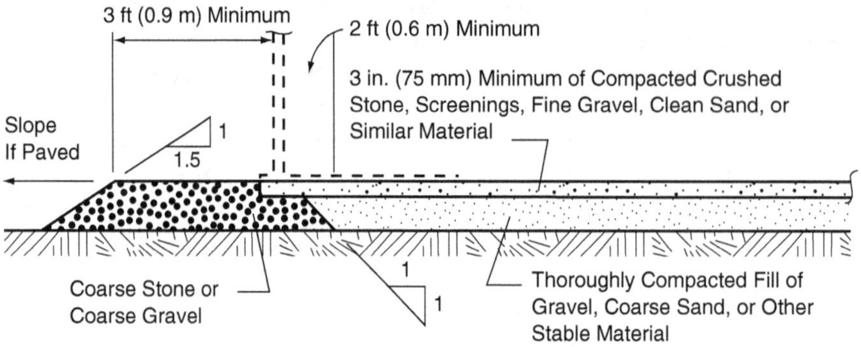

NOTE: Bottom of base excavation should be level. Remove any unsuitable material and replace with suitable fill, then thoroughly compact same.

Figure 6-2 Example of tank supported on granular berm foundation

Foundation types 1, 3, 4, and 5 use a fine-grade aggregate (usually clean sand) that is placed immediately under the tank bottom. For many years this sand cushion was saturated with light oil or diesel fuel to add cohesive properties and to serve as a corrosion deterrent. Regulatory agencies have been concerned that the oil from the sand might leach into the earth beyond the tank foundation. Where this potential leaching is a concern, the purity of the sand should be controlled and hydrated lime may be added to enhance the corrosion resistance of the sand cushion. This modification of the sand cushion is discussed in ANSI/AWWA D100.

Granular Berm Foundation

If the tank need not be anchored to resist wind or earthquake forces, the most economical type of foundation is a granular berm. This type of foundation (Figure 6-2) has been used successfully for many years.

When the unrestrained granular berm foundation is chosen, attention must be paid to the choice of berm material and drainage around the tank to ensure against foundation washout. A 1-in. (25-mm) rainfall on a tank of a radius of *R* ft (m) will produce a sheet of water 1 in. (25 mm) wide and *R*/2 ft (m) high washing down the sides of the tank. Therefore, any type of flat-bottom tank foundation must be designed carefully. Steel retainer rings may be installed to ensure berm stability.

Ringwall Foundation

The most common type of water-reservoir foundation is one for which the tank shell rests on a concrete ringwall and the tank bottom rests on well-drained, noncorrosive, compacted fill. This type of foundation does not have the tendency to wash out like the granular berm foundation. Tanks may be placed within the ringwall. If settlement occurs within the ringwall, water may be trapped within the ringwall, causing a corrosive environment for the tank's lower shell and bottom. Therefore, provision for proper drainage from inside the ringwall is necessary.

Concrete Slabs

Tall standpipes or ground storage reservoirs erected on unstable soils may be erected on concrete slab foundations. The concrete, if sloped and well drained, provides an alkaline environment that normally mitigates corrosion on the underside of the tank bottom.

Bolted Tanks With Concrete Bottoms

In addition to the foundation types previously discussed, ANSI/AWWA D103 provides for an alternative tank design: bolted steel tank with the sidewall embedded in concrete (foundation type 6). In this type of tank design, a reinforced concrete slab serves directly as the tank bottom, replacing the welded steel plate used in other tank designs, and the steel sidewall is embedded in the reinforced concrete bottom slab. The foundation for this type of tank construction consists of a concrete ringwall and interior granular base course over which the reinforced concrete bottom slab is laid. For each site, specifications for the preparation, installation, and compaction of the subbase material should be obtained from, or developed in consultation with, the tank manufacturer.

ELEVATED TANK FOUNDATIONS

The combined steel, water, seismic, and wind loadings of elevated tanks are carried to the earth by individual column piers, circular ringwalls, or a combination of these, depending on the design and configuration of the column support system. The piers or ringwalls usually interface directly with the earth at the depth prescribed by the geotechnical engineer. The sizes of the slabs or pads beneath the piers or ringwall are functions of the loadings applied to the column and the safe supporting value of the soil. In poor soils, pilings or caissons may support the piers or ringwall. Sometimes soil replacement or consolidation of on-site materials is necessary.

TANK SITE

The following listing provides criteria to consider for selecting a tank site:

1. Is the site accessible on public roads by a semitrailer tractor rig with a 40–45-ft (12–13.7-m) long trailer requiring a vertical clearance of 14 ft (4.3 m) and weigh ing 73,200 lb (33,200 kg) gross weight?
2. Is there an access road or easement to the site? If there is only an easement, who will pay for the road? Will the access road have to be taken out at the end of the project? Can pipes to and from the tank be located in this access route?
3. Are there power lines or other overhead obstructions over the site, over the proposed access road, or beside the site that will interfere with safety during construction, coating, or maintenance operations? Will a power line be closer than 40 ft (12 m) to any point on the tank?
4. Are there underground obstructions on the site such as gas lines, sewers, or buried electrical or telephone cables? Was this a burial site?
5. Does the site have or can it be made to have good drainage for ease of construction operations, tank maintenance, and protection from foundation soils saturation? Is there provision for draining the tank?
6. Is the earth firm enough to support construction equipment during "normal" weather conditions?
7. Is the site large enough for
   a. erection equipment, steel storage, staging operations, ground assembly operations, and a safe "fallout area" for items dropped from the tank during erection?
   b. maintenance of the tank and piping after completion?

c. abrasive blasting and coating of the tank during initial construction and during future recoating?

8. Will pile driving disturb or cause failure or damage to neighboring foundations or superstructures?
9. Will the noise from pile driving, excavation, steel erection, or abrasive blasting operations be unacceptable to neighboring property use such as a school, hospital, or nursing home?
10. Will the tank obstruct concerned citizens' views of natural vistas, historical landmarks, and so on?
11. Will the construction activity be a local-area hazard to traffic safety or youngsters in after-hours play? Can the site be made secure during construction by fencing?
12. Are there trees or other vegetation on the site that are required to be preserved as a condition of purchase or use that would interfere with efficient, safe construction on the site? Will the site soils support the growth of trees planted to screen the site?
13. Can three-phase 480-V power be made available on the site for construction operations? (This is not required, but it frequently lowers construction costs.)
14. Is the site location convenient for hookup to water mains, telemetry wiring, and permanent electrical power?
15. Will the tank be in an area subject to vandalism?
16. Has the site been subject to an environmental assessment prior to acquisition or use?

Location

The likelihood that a proposed tank site will be marginal or unacceptable increases as the cost of land increases, as the availability of land declines, and as public opinion places more stringent requirements on the location of water-storage facilities. The forewords of ANSI/AWWA D100, ANSI/AWWA D103, and ANSI/AWWA D107 require the owner to provide the site. This site should be graded to accommodate the proposed tank and should have sufficient room to allow the structure to be built by customary methods. The site should be located away from buildings, heavily traveled public thoroughfares, parking lots, and livestock, at a distance sufficient to limit potential damage or injury to third parties resulting from construction activities. The site should also take advantage of the local terrain. However, in high-vandalism areas, remote locations tend to be a great source of concern and expense to owners. The tank site should be free of all overhead and underground obstructions and utility lines. Power lines located on the perimeter of the site must be a minimum of 40 ft (12.2 m) from the edge of the tank or the tank foundation, whichever is closer to the power lines.

Size

Final property lines should be located sufficiently far from the foundation to facilitate maintenance of the tank and site. At a minimum, the outer extremities of the foundation should be at least 20 ft (6.1 m) inside the permanent tank site boundaries. Additional clearance will be necessary for construction operations and tank component storage and subassembly. If a permanent site of adequate size is not available, the owner should provide an adequate temporary construction easement.

With elevated tanks and standpipes it is preferable, when the space is available, to have a site clearance from the center of the tank to the site extremities equal to the height of the tank.

Drainage

The tank site should have good drainage. A site with good drainage will minimize or prevent possible foundation subsidence due to the long-term intrusion of surface water. It will also minimize possible corrosive conditions under the tank bottom and will probably result in lower construction costs. The site design should also include provisions for draining the tank contents and the discharge from the tank overflow without damaging the tank site or neighboring properties.

Access

The owner should provide a suitable right-of-way for access from the nearest public road to the tank site. The access should be able to handle tandem tractor trailers with legal loads under ordinary weather conditions, and it should be free of underground and overhead obstructions that would be damaged by the traffic. If the right-of-way is permanent, a permanent roadway should be installed once the tank project is completed.

Other Siting Considerations

Since construction and maintenance of tall tanks involve procedures that may disturb neighboring property, it is frequently advantageous to locate the tank on a relatively large area of undeveloped land, dedicated open space, or greenbelt. A minimum clear distance of approximately 100 yd (90 m) between the tank and the open space boundaries will greatly reduce future maintenance costs.

When tanks are to be constructed in conjunction with other water utility facilities such as a treatment plant, the problems associated with scheduling the simultaneous operations of multiple constructors must be considered during the tank site selection process.

TANK COATING: WELDED STEEL TANKS

Chapter 4 discusses tank coating in detail. The following material relates to the specification of interior and exterior coating of newly constructed tanks. The decision to shop clean and prime or totally field-clean and coat should be based on the size and configuration of the tank, the tank site location, environmental regulations, and the climate anticipated during the cleaning and coating processes.

Selection of Coating

Initial coating of field-welded tanks represents approximately 10 to 15 percent of the total cost of a new tank (including foundations). Two-coat systems are used on the tank; both coats are critical. Foremost in the eyes of the community is the exterior finish color and its aesthetic appeal. However, the interior surface (which few see) is far more critical; the interior is where the most aggressive and detrimental conditions of corrosion can occur undetected.

The coating system should be selected based on information provided by tank and coating constructors and the coating manufacturer's representatives, local ordinances, the engineer's preference, and the owner's preference.

Interior Coating

An interior coating in the lower areas of the tank must withstand constant immersion; it must be able to resist alternate wetting and drying in the upper portion of the operating range and high humidity above the high water level; it must be resistant to the actions of ice abrasion in cold climates; and it must be able to adequately withstand local temperature extremes and water characteristics. The system must be receptive to the conditions of application (primarily temperature and humidity) in order to be cost-effective.

Interior coating systems must meet the minimum requirements of NSF/ANSI Standard 61 with regard to acceptable extraction levels of the constituent materials.

Inspection. The integrity of the tank's inside coating system is only as good as the quality of coating materials, cleaning, application, and curing. Adequate inspection during construction is necessary to confirm the quality of cleaning and coating operations. Inspection of the interior surfaces helps to detect minor coating deficiencies, which can then be corrected, preventing the propagation of coating failures.

The maximum interval for periodic inspection of the tank interior should normally be 3 years. It is usually advisable to wash out the tank at the time of inspection. The owner should keep accurate records of the inspection and maintenance functions performed on the tank. This record should include copies of specifications, contracts, and inspection reports and coating manufacturers' product numbers and product data sheets.

Rafter-supported roofs. For maximum corrosion protection, roof support beams and their intersections with the roof plates on ground storage tanks should be shop- or field-coated prior to erection; otherwise, these areas will not be protected from corrosion. Although coating of these areas is highly desirable, it is not standard industry practice. Therefore, the purchaser must specify this precoating requirement in the bid documents; otherwise coating prior to erection will not normally be performed.

Exterior Coating

Exterior coating systems normally attract more attention and cause more concern than any other aspect of a tank project. Prior to 1980, alkyds and silicone alkyds were most commonly used on the exteriors of new tanks. Since then, urethane-based coating systems are found on an increasing number of tank exteriors. The alkyds were the most popular coating system in the past because they were easy to apply and generally accepted by owners. The silicone alkyd materials have better gloss and color retention than the straight alkyd materials. A purchaser desiring a bright color may select a urethane system for its superior color and gloss retention; however, additional costs and site restraints are associated with these materials. The tank location plays an important part in the selection of an exterior coating system, since some materials (particularly the urethanes and long-oil alkyds) tend to produce more overspray and damage to surrounding property than other materials.

Artwork

Decorative artwork and lettering schemes can be added to the tank within the constraints of the tank's geometry. A rule of thumb is that a tank sign or logo should be no longer than three-fourths of the tank diameter for it to be viewed from one vantage point. The maximum height of a tank sign should also be limited, if possible, so that the letters in the sign do not arch around the curved upper portion of the tank and appear distorted.

TANK COATING: BOLTED STEEL TANKS

ANSI/AWWA D103 requires that tank parts receive a near-white (SSPC-SP10) blast or, as an alternative when fused-glass coatings are used, pickling (SSPC-SP8) and a coating by one of the systems listed in Section 12 of that standard. The standard also requires that the critical interior surface receive a USEPA-accepted, amine-cured thermoset liquid or powder epoxy coating; a fused-glass coating; or galvanizing. This requirement helps ensure that the tank is able to provide safe, long-term storage under a wide range of temperatures, humidity conditions, and water types.

Aesthetically pleasing, long-lasting performance is a major consideration in choosing an exterior coating system. Either (1) an epoxy primer with a baked acrylic or acrylic-urethane topcoat; (2) a thermoset powder coating of epoxy, acrylic, urethane, or a hybrid; or (3) a fused-glass coating can provide this performance. Galvanizing is also an option for exterior corrosion protection.

TANK WATER TESTING AND DISINFECTION

Concurrent Versus Separate

Tank disinfection procedures are normally performed in accordance with ANSI/AWWA C652 or the more stringent requirements of local health agencies. ANSI/AWWA D100 requires that the tank be tested prior to coating and that the owner furnish the water to fill the tank and provide a means of disposal following testing. ANSI/AWWA D100 permits the coating to be performed before water-testing the tank if specifically agreed on between the tank constructor and the purchaser. Water testing and disinfection phases of the project could then be concurrent, thus saving the owner the cost of producing and disposing of a large quantity of water. However, water testing the welded tank prior to coating has the following advantages:

- It allows identification and correction of any distortions caused by anticipated or unanticipated foundation settlements prior to coating.
- It allows the identification and correction of any leaks that might have been temporarily covered by the interior coating system.

Valves and piping should be tested in accordance with ANSI/AWWA C600.

The constructor may be required by contract to provide and dispose of the testing and disinfection water. However, the purchaser will ultimately bear these costs, plus the constructor's overhead and profit.

ENGINEER'S ROLE

Water-storage tanks are generally classified as engineered construction, in which the engineer is not a party to the contract between the owner and constructor but rather is employed by the owner to prepare the construction plans and contract documents, interpret the contract, judge performance, and observe the construction. Under the contract between the engineer and the owner, the engineer agrees to provide a specified set of plans and specifications and a package of services. The drawings included in the plans show the extent and arrangement of the components on the project. Tank projects generally require a limited number of drawings, usually consisting of an elevation drawing, an orientation of the tank accessories, and detailed information regarding site access and related site work. The forewords of ANSI/AWWA D100, ANSI/AWWA D103, and ANSI/AWWA D107 each contain a list of information to be provided by the owner or engineer concerning elevated tanks, standpipes, or reservoirs.

Accessory Detail

Accessory items on the tank structure should be located where they are readily accessible from a fixed ladder or platform surface. An exception to this is the location of stub overflows, which when used are purposely located away from ladders to avoid ladder icing. Specific tank accessories required by AWWA standards should be shown but not detailed on the bid drawings, since each manufacturer has proprietary components that fulfill the intent of the standards. Such details may cause problems if bidders are required to provide another constructor's proprietary apparatus.

Accessory details and orientation should be developed and included in the shop drawing submittals after the contract is awarded. The governing requirement should be that the accessories meet the minimum requirements of the referenced AWWA standards and the intent of the specifications. If the owner elects to include apparatuses and operating systems in excess of those specified by AWWA standards, the engineer should provide specifications and details that clearly define the components required and the scope of related work. A more thorough description of appurtenances and accessories on steel tanks is given in chapter 3.

Federal Aviation Administration

The owner or his or her engineer should file notification of construction with the Federal Aviation Administration (FAA) prior to construction of tall standpipes or elevated tanks, in order to determine the requirements for temporary and permanent tank markings and lighting.

BIDDING DOCUMENTS

This section discusses documentation and other issues related to the bidding process.

Technical Specifications

The project specifications describe the materials and work required to construct the tank. These documents are a basis for bidding and negotiation, and they serve as instruments by which the contract may be administered.

Because the initial purpose of the plans and specifications is to inform prospective bidders about the project, these documents should help the constructor become familiar with local conditions and local building requirements. Competitive bidding is effective only when the project plans and specifications are complete and accurate and when duly qualified constructors are given sufficient time to bid on the project. The engineer can obtain valuable assistance from constructors and tank manufacturers when writing specifications.

When incorporating the requirements of ANSI/AWWA D100, ANSI/AWWA D103, or ANSI/AWWA D107 into the project specifications, the engineer is working on the premise of a design type specification that describes the kinds of materials to be provided by established standards and ensures performance by specified inspections and tests. It should be noted that ANSI/AWWA D103 contains coatings requirements for factory-coated tanks.

Alternative Bid Items

In preparing the bidding documents, the purchaser must carefully evaluate the needs of the system and aesthetic considerations. Ultimately, the type of tank that will best satisfy the owner's needs will be determined primarily by economic considerations, since many designs can equally satisfy the intended function of storing some quantity of water at a given elevation. To help determine the best tank value for the money, the owner may consider specifying certain features as alternative bid items. In the best interest of the owner, the engineer, and the constructors, only realistic alternative bid items should be considered.

Documents Related to Used Tanks

One potential advantage of steel tanks is that they can be relocated or possibly sold if the level of system demand changes. However, caution should be exercised in evaluating a used tank that is bid as an alternative to a new tank. The initial cost should not be the only consideration. Total life cycle maintenance costs should be projected. If a used tank is considered, the owner should require (1) a copy of an up-to-date inspection report of the structure provided by a qualified professional engi neer, (2) a signed document from the present owner stating that the tank is available for sale, and (3) a detailed proposal of any remedial coating or repair work that will be done to bring the structure to the equivalent state of the proposed new tank. It is also important to know the specifications and standards under which the tank was originally built, as well as the wind and seismic loadings for which it was designed and constructed. One additional caution concerns the possibility of lead or other hazardous materials being in the coating of the tank being considered for reuse.

It should be noted that used tanks designed and used for fuel oil or other non-potable water use will likely have been designed to contain a less dense liquid and consequently will not be of adequate thickness for water storage. Additional problems in converting a tank from hydro storage into potable water use have to do with the cleanliness of the steel that can be achieved after such service and the effect this will have on the adhesion and the purity of the coatings applied to the water contact surfaces.

Before purchasing a tank that has previously been used for purposes other than potable water storage, check with local, state or provincial, and federal authorities, as many authorities prohibit such reuse.

Foundation Unit Prices

The contract price of a tank structure of specified size and design is normally well defined and is not subject to wide variations as construction progresses. Foundation costs, however, tend to be difficult to estimate accurately and may be subject to change because of possible variations in subsurface conditions. Bidding documents should recognize this fact and provide a means to adjust work and material quantities to meet the conditions found during construction. If the engineer anticipates changes, then he or she should provide information regarding any additional labor or materials that may be required. Most constructors prefer an add/deduct unit-price adjustment to be an integral part of the bid. The basis for the bid should be the constructor's design, based on adequate information provided by the project plans and specifications. For example, pile foundations have sometimes been bid on a lump-sum basis. However, it is difficult or impossible to determine the exact length of piling required for a given site. Various factors influence the actual length of the piling, and a fixed-price lump-sum contract can create a difficult and inequitable situation for either the constructor or the owner.

Constructor Assistance

Generally, tank constructors and manufacturers are more than willing to discuss individual tankage needs and to assist the owner and the engineer by providing standard design information for a project. For an elevated tank of a given capacity, each constructor and manufacturer will have different geometric parameters. These dimensions normally do not vary enough to cause difficulty on a system if one constructor is selected for a project based on the information supplied by another constructor.

The purchaser is advised to contact prospective bidders and discuss a project prior to issuing an invitation to bid, in order to familiarize the purchaser with current industry standards and practices. Most constructors and manufacturers are willing to provide copies of preliminary specifications that have been developed for water tanks of different styles and capacities. The purchaser must be careful to make every effort to write a specification that is open and does not exclude bidding by any qualified manufacturer or supplier. In particular, a given manufacturer's proprietary design details should not be included in a project's contract documents; this would create an inequitable bidding situation for other qualified suppliers or manufacturers.

Chapter 7

Inspecting New Tank Construction

The purchaser should inspect the tank's construction as it proceeds to ensure that the structure complies with specifications, both in form and in quality. This chapter discusses specific items to be checked by the purchaser and items that are the responsibility of the constructor.

On the job, the purchaser and constructor should keep in mind that a quality structure is the result of the cooperation and affirmative efforts of all parties. If the relationship between the purchaser and constructor is combative or antagonistic, the project is likely to be completed late, with a quality that is adequate at best. However, a purchaser with a sound knowledge of specifications, standards, and trade practices can work with the constructor to complete a high-quality project on schedule.

RESPONSIBILITY FOR QUALITY

In general, tank constructors are responsible for maintaining the quality of their work. ANSI/AWWA D100, Standard for Welded Carbon Steel Tanks for Water Storage, ANSI/AWWA D103, Standard for Factory-Coated Bolted Carbon Steel Tanks for Water Storage, and ANSI/AWWA D107, Standard for Composite Elevated Tanks for Water Storage, give the constructor the responsibility for designing and conducting a welding or bolting quality program. According to the provisions of ANSI/AWWA D103, the constructor is responsible for selecting a factory-coated bolted tank (manufactured in accordance with that same standard) given the capacity and height or diameter requirements of the owner. This is a logical assignment of responsibility, because the designing and building of tanks is a specialized field and requires specialized equipment and knowledge to conduct a total quality control program. Nonetheless, the owner or the owner's engineer, as well as third-party consultants where needed, should monitor the work for inevitable human errors and misunderstandings.

Even before construction begins, shop drawings should be checked by the owner or engineer to ensure that the special requirements of the job are met. The constructor is usually responsible for the structural design of the tank, but the interplay between the legally defined responsibilities of the engineering and construction phases is not firmly established. The engineer reviewing the drawings should clearly specify the purpose and the extent of the review to avoid being held legally liable for a more detailed examination than was actually performed.

A quality assurance program for the tank project should take into account each of the functions involved in building a tank. In general, these functions fall under the following categories: foundation, fabrication, steel delivery, tank erection, field-applied coatings, shop-applied coatings, and appurtenances.

THE FOUNDATION

Most owners and engineers apply greater expertise to inspecting the foundations than to any other function of tank construction. If the owner or his or her engineer does not have expertise in this area, other experts should be contacted. The foundation for a tank is the most significant source of potential failure. The reinforced concrete foundation and the soil on which it bears are made up of nonhomogeneous materials, over which there is little control. In addition, the foundation is frequently installed by a specialty constructor or under another division of the water project, giving the tank constructor little control over this function. Figure 7-1 shows foundation construction in progress.

Soil Investigation

The purchaser should be at the site when the borings are being taken as part of the soil investigation. This gives the purchaser better insight into problems that might be encountered during construction of the foundation.

Activities Before Concrete Is Placed

The soil conditions at the bottom of each excavation should be evaluated to determine if they are the same as those used when the foundation design was developed. If piles are being driven, the purchaser must determine if the piles are driving as predicted. The purchaser should be on the job to verify the pile-driving log. Site and final concrete elevations should be confirmed. Placement of forms and reinforcing steel should be verified and photographs should be taken to document this before the concrete is allowed to be placed.

Concrete

If the owner, constructor, or engineer does not have extensive previous experience with the materials supplied by the concrete ready-mix plant, a design mix should be developed and tested. The consistency of the concrete should be evaluated as it comes out of the chute, and concrete test cylinders should be taken. The anchor-bolt placement should be checked, as well as the pier elevation(s). Tolerances on concrete foundations are given in ANSI/AWWA D100, ANSI/AWWA D103, and ANSI/AWWA D107. The purchaser's representative should be familiar with American Concrete Institute Standard ACI 301, Specifications for Structural Concrete.

Quality concrete with a good cure will ensure the compressive strength that the concrete must develop and will help prevent surface deterioration. If the surface of the foundation is not cured properly to protect it from the elements, a chipping or spalling failure of the concrete can occur long before the steel tank deteriorates.

Figure 7-1 Tank foundation construction

Backfilling

Proper backfilling techniques are necessary to provide a structurally stable foundation, to prevent overstressing of the concrete during backfilling, to prevent water ponding on moisture-weakened soils, and to make the site more solid for the tank-erection crew.

Proper backfilling operations require the specialized services of a qualified soil-testing laboratory to (1) determine the optimal moisture content and maximum density of the backfill material, (2) check the moisture content of the material being placed, and (3) conduct relative density tests in the field after the backfill material has been placed and compacted to ensure that the specified degree of soil compaction has been obtained. Adequate soils compaction is particularly important for the foundations of ground storage tanks. Although the bottoms of these tanks are usually quite flexible, particular care is necessary for backfilling pipe trenches beneath the tank and the soil or fill material adjacent to concrete ringwalls, where severe differential settlement can cause rupture of the underlying piping and possible failure of the tank bottom.

The contract documents may assign the responsibility for providing necessary soil-testing services to either the tank constructor or the owner. In either case, copies of all soil test reports should be furnished promptly to all interested parties.

FABRICATION

It is recommended that the owner and engineer visit the constructor's fabricating facility while the tank is being fabricated. They should inspect the quality of shop fabrication, the type of surface preparation, and the shop coating if applied. This visit also serves to open communications for the balance of the project.

STEEL DELIVERY

The owner should be on hand when the steel is delivered to the job site. The owner can help resolve conflicts with neighboring property owners, document any damage occurring in the unloading process, and protect underground utilities on the site or under the access road.

TANK ERECTION

Erecting and welding or bolting the steel are tasks for which the expertise of the constructor is vital to the success of the project. Erecting steel is a dangerous operation, requiring skills acquired only through experience. During this phase, the purchaser's representative may need assistance. Independent testing laboratories are usually equipped to take radiographs of welded seams, but they know little about steel erection and fit-up and are not willing to climb the heights usually associated with water-storage facilities.

Using someone from another tank constructor's organization as the purchaser's representative can lead to conflicts of interest and other problems. It is very difficult for a competitor to be unbiased in the evaluation of another constructor's work. Even if this competitor is fair, it is difficult for the tank constructor to accept the opinion as an unbiased one. Therefore, it is usually best to secure the services of a consultant who specializes in this type of inspection work and has the expertise and climbing ability to accomplish the job.

Fit-up Quality

The levelness of the tank's base plate(s) is critical if the rest of the tank is to be erected properly. The constructor's steel erection supervisor should check the foundation(s) for differences in elevation. Any such differences should be compensated for by shims underneath the base plate(s). If this task is not done properly, or if there were fabrication problems with the steel, the purchaser may see slivers of steel that need to be cut from seams, frequent use of a large hammer to form the steel, variations in the seam gaps, or plates not aligned in accordance with the tolerances required in ANSI/AWWA D100, ANSI/AWWA D103, and ANSI/AWWA D107. These problems usually produce a tank of unacceptable aesthetic or structural quality.

Welding Quality

According to ANSI/AWWA D100, ANSI/AWWA D103, and ANSI/AWWA D107, the constructor is required to check the quality of the welding. It is the job of the owner to monitor the constructor's quality control program. For welded butt joints, the most common method of evaluating weld quality is by means of radiography. The purchaser should participate in the selection of radiograph locations, watch for documentation of the radiographs, and review the radiographs with the constructor's quality assurance expert. ANSI/AWWA D100 and ANSI/AWWA D107 have radiograph inspection standards that must be used for evaluating discontinuities and defects present in radiograph film. It is vital for the tank constructor and purchaser to visually inspect all welds to ensure the removal of all weld splatter, sharp surfaces, overlaps, and unaccep-

Figure 7-2 Typical welding operation in the field

able undercuts that will be detrimental to the coating life. Welds do not need to be perfectly smooth, but they do need to have sharp edges removed. Ground storage tanks erected under the Alternative Design Basis (Sec. 14 of ANSI/AWWA D100) require many more radiographs than standard tanks. Figure 7-2 shows a typical welding operation in the field, Figure 7-3 shows a weld radiograph review, and Figure 7-4 shows an almost completely erected elevated tank.

Bolting Assembly

Bolted steel tanks require the proper placing of steel sheets, gaskets, and sealants. Some erection methods may also require pretensioning the sheets and tightening the bolts to a prescribed torque. These details are covered by the manufacturer's erection instructions and drawings. The engineer or purchaser may require that a set of these instructions be included with the shop drawing package that is submitted.

Tank Appearance

Tank appearance is of great importance to many owners. Problems with appearance are difficult for the purchaser to notice until it is too late because the final appearance often is not known until the tank is coated. It is then that the dents and buckles become apparent and the owner expresses dissatisfaction. Determining how well the tank complies with the specifications and applicable codes and negotiating a settlement for poor appearance is a time-consuming and stressful procedure. Usually these problems will be avoided if the constructor checks to see that the tank is level, round, and plumb as it is being built. Incorporating dimensional tolerances into the contract will also minimize disputes. ANSI/AWWA D100, ANSI/AWWA D103, and ANSI/AWWA D107 incorporate some of these tolerances.

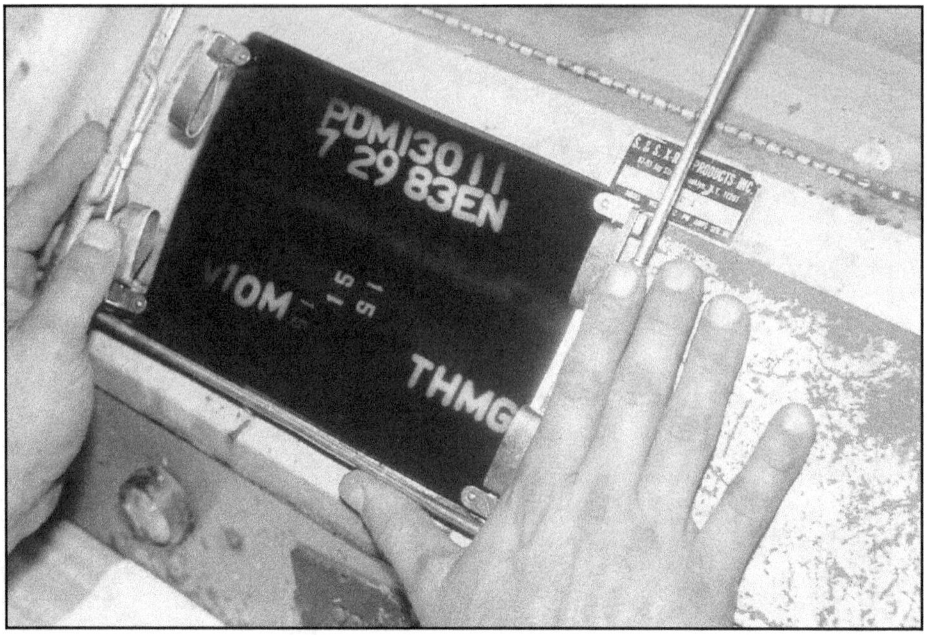

Figure 7-3 Reviewing a weld radiograph

Surface Regularity

A smooth, regular surface will provide a good basis for the application of a protective coating system, thus helping to reduce maintenance costs. To this end, the constructor and the purchaser should ensure that the weld contour is smooth, that unacceptable weld undercutting is eliminated, that weld spatter is ground off, that remains of welds used to attach erection and fit-up equipment are chipped and ground smooth, and that unacceptable gouged-out places in the steel are filled in.

The tank constructor and purchaser's representative should be alert to find all sharp edges or areas that would cause premature coating failure so that action may be taken as the work progresses.

Factory-Coated Bolted Tanks

A bolted steel tank is delivered to the location with a factory-applied coating. If the steel has not been damaged in transit, the surfaces will be smooth. Each panel should be carefully inspected before erection. If there has been any damage in transit, in handling, or in erecting the tank, the panel may need to be repaired or replaced.

Water Testing

When welded tanks are water tested before they are coated, any leaks that are found can be repaired without requiring any coating to be redone. If the tank is not filled until after coating, small pinholes in the welds may be plugged temporarily with coating; these will cause leaks later if the coating breaks loose. The owner should ensure that water for the test is available at the time and pressure necessary to coincide with the constructor's schedule. The owner should also ensure that provisions are made for draining and disposing of the test water. If leaks are found in factory-coated bolted tanks, the constructor should make repairs according to the manufacturer's recommendations.

Figure 7-4 Newly erected elevated tank

FIELD CLEANING AND COATING

This section discusses the cleaning and coating of welded steel tanks after they have been erected and before they are placed in service.

Erection Scar Removal

By the time the tank is erected, irregularities in the surfaces of the tank should have been eliminated. The erection crew has the equipment and scaffolding to smooth out these defects. The coating crew is often subcontracted or is comprised of members of another department of the tank company. Their rigging is sometimes not as well suited for this as the scaffolding of the erection crew; therefore, they cannot perform this cleanup work as economically or effectively.

Occasionally, small laminations in the steel plate will be revealed by abrasive blasting. If these are not removed completely by blasting and remain large enough to produce holidays in the coating system, they should be removed by grinding. Occasionally, deeper laminations may require welding or further testing.

Steel Cleanliness

The first requirement for a good coating is a clean surface. The steel should be free from dirt and oil, both of which may accumulate during construction. All weld seams, abraded areas, scratches, shop or field markings, or poorly adhering shop primer should be removed by abrasive blast cleaning. The areas cleaned by abrasive blasting should blend well into the adjoining undisturbed shop primer. Some shop-applied primers must be scarified or otherwise prepared before ensuing coat(s) are applied.

The purchaser should also be aware that welding or cutting activity on one side of a plate will likely damage the coating bonding on the opposite side of the plate. This is especially important if shop priming is used. The areas opposite welding or cutting operations should be examined for coating damage resulting from the heat induced by the cutting or welding process.

The manuals *Good Painting Practice* and *Systems and Specifications*, visual standards, and inspection manuals available from the Society for Protective Coatings (SSPC) all provide good guidelines for inspecting coating.

Inspection Instruments

Instruments needed to inspect coating include at a minimum a wet-film thickness gauge, a calibrated dry-film thickness gauge, equipment for measuring air temperature and humidity, a steel-temperature thermometer, a surface-profile measuring device, and a wet-sponge-type holiday detector. The holiday detector is used to inspect the coating for voids that will cause premature coating failure. If full-time inspection is not conducted, a Tooke Gage or other destructive-type testing will be necessary to evaluate the thickness of each coat and obtain an indication of the cleanliness of the substrate.

Inspection Planning

The purchaser should plan work to aid in the timely completion of the tank field coating. This will require open lines of communication with the coating company and an understanding of the effects of weather on coating progress. The constructor will also need to work efficiently in good weather.

The purchaser should state requirements for the number of locations to be tested according to the total plate surface area. Minimizing testing is unwise, but an excessive number of testing locations places an unreasonable burden on the constructor and can substantially delay the progress of the tank coating. SSPC PA-2 delineates procedures for measuring coating dry film thicknesses. Large tanks may require more than one purchaser's representative to conduct the required field tests in order to avoid excessively delaying the coating progress.

Technical Aspects of Coatings

Today's coatings require exactness in measuring and mixing components and thinners. The appropriate application equipment must be used, and the proper combination of humidity and dew point, air temperature, and steel temperature are critical during both application and curing.

The tank interior must be ventilated to ensure the safety of workers and the proper curing of the field-applied coatings. Fans or air horns are usually required to move air through the tank. Even with forced-air ventilation, proper breathing equipment is necessary for the safety of the workers and the purchaser's representative(s).

Bolted tank panels are coated at the factory under controlled conditions. ANSI/AWWA D103 requires that the panels be grit-blasted to near-white metal (SSPC SP10) and coated within 15 minutes of cleaning to prevent rust from starting. The coating is then either baked on or fused on. These operations may be observed by the purchaser at the time of the shop inspection if desired.

MECHANICAL AND ELECTRICAL APPURTENANCES

The mechanical and electrical items that need to be checked will vary with the equipment specified for a particular project. The following should be checked on every project:

- Electrical wiring should meet applicable codes.
- Conduits, fixtures, pipes, valves, or other items should not interfere with the safety of the ladders or platforms.
- Cathodic protection anodes and hand-hole covers should be properly placed, and the purchaser's representative should witness the potential profile being conducted.
- All hatches should be locked.
- Safety belts and sleeves should be furnished for the ladder safety devices. Safety sleeves should be checked for proper operation along the full height of the rails or cables. Any coating, deviations, or obstructions that prevent the free operation of the sleeve should be removed.

This page intentionally blank.

Part III

Existing Tanks

Routine Operation and Maintenance

Professional Examination and Renovation

Cold-Weather Operation

This page intentionally blank.

Chapter 8

Routine Operation and Maintenance

Tanks need to be operated properly and maintained and inspected on a regular basis to maintain their structural and sanitary integrity. This chapter covers routine operation and maintenance performed by water system operators, which represent a crucial complement to the professional tank inspections discussed in chapter 9. Routine tank operation typically consists of the daily monitoring of the automated systems designed to level out supply and demand. Tanks fill when supply exceeds demand and empty into the system when demand exceeds supply, typically maintaining water pressure between 35 and 100 psi (240 and 690 kPa). However, many other factors must be taken into account in the day-to-day operation of a water distribution system.

Another important aspect of tank operation is the prevention of freezing problems in tanks. For a complete description of freezing problems and their solutions, see chapter 10.

ENERGY MANAGEMENT

A significant amount of energy can be saved by proper management of the distribution system, including storage. Since pumps are generally more efficient when they pump against lower head levels, it is more efficient to allow tank levels to fluctuate rather than to always keep the tanks full. A water utility can also achieve significant savings by using the most efficient pumps first and by replacing worn parts that may decrease the pump efficiency. Variable-rate pumps should be used at the highest possible speed, at which their efficiencies are usually greatest.

Another consideration involves maximizing energy cost savings by avoiding peak rate charges. If the storage capacity is sufficient, the pumping times can be modified to take advantage of off-peak energy charges. Chapter 5 includes a further discussion of this cost-saving factor.

CONTROLS

The controls used in automated distribution systems require periodic inspection and maintenance to ensure trouble-free operation. Without regular inspection of the controls, potential breakdowns would not be discovered and more emergency repairs would be required.

Altitude Valves

Altitude valves are used to maintain a range of acceptable water levels in the tank. The control mechanism on an altitude valve is called a pilot valve.

Maintenance of the altitude valve should be conducted as the valve manufacturer recommends, and additional maintenance of the altitude valve should include the following:

- The pilot valve should be tested and lubricated.
- Valves should be exercised. Planned exercising of valves keeps them clean and operable, extends their life, and pinpoints problem valves, allowing repairs or replacements to be scheduled before an emergency occurs.
- It is important to check for leakage, keeping the valve port open at all times; leakage at this location indicates a worn piston face or a scored piston or liner surface. Constant leakage from pilot valve waste lines indicates a defective pressure valve, which must be reground or replaced. Leakage through the main valve indicates worn seatings.

More involved maintenance of the altitude valve, including dismantling and examining the piston and cylinder walls, should be performed if any evidence of wear is found. The valve manufacturer should be contacted for maintenance and repair procedures.

PERIODIC OPERATOR INSPECTION

Although it is important that the periodic professional inspections described in chapter 9 be performed on tanks, some critical portions of the tanks should be inspected by the water system operators. The factors in the following paragraphs should be checked at least once a month, or weekly if possible; however, any elevated inspection should be conducted only by experienced climbers equipped with the proper safety equipment.

Foundations

The foundations and surrounding earth should be examined for any signs of settlement. The concrete should also be observed for crumbling, deep cracking, and exposed reinforcing steel. If any of these conditions is found, the tank should be professionally inspected at the earliest opportunity.

Leaks

The exterior of the water-bearing surfaces should be examined, and any leaks—or rust streaking that could have been caused by tank leaks—should be reported. (Corrosion products often seal leaks, leaving only rust streaks as evidence of the leak.) The tank should be inspected by a professional structural engineer familiar with water tank construction as soon as possible after the leak is discovered. Although some leaks may not cause structural problems, potential catastrophic tank failures can be avoided if the visible leaks are properly investigated and repaired.

Cathodic Protection

If the tank has been equipped with an automatic impressed-current cathodic protection system, the system's supplier should inform the tank owner of what meter readings are acceptable. The operator should check these meters and inform the supplier of any significant changes.

Exterior Corrosion

Any exterior corrosion, especially where metal loss is apparent, should be evaluated by a professional engineer familiar with the construction of water-storage tanks. If the operator notices a change or severe worsening of the exterior corrosion patterns, he or she should bring this to the attention of the engineer. Special areas to observe are anchor bolts and nuts, rods, and rod pins and clevises.

Vandalism

The locks on ladders and access doors should be checked to prevent vandalism.

Ladders, Platforms, and Lighting

As the tank is accessed, the ladders and any ladder platforms should be inspected for noticeable metal loss. Any such metal loss should be inspected professionally.

If the ladder extends up an "interior dry" area, such as is found on single-pedestal elevated tanks and on some large-capacity multiple-column tanks, the access door to the area should be kept locked, and the interior dry area should be sufficiently lighted. Broken or missing lightbulbs should be replaced.

If the ladder extends up the exterior of the tank, the ladder should be equipped with a locked guard to prevent unauthorized access to the tank exterior and roof.

Overflow, Manholes, and Vents

The items that most directly affect how sanitary the tank is are the overflow, manholes, and vents. Oftentimes contamination in a public water system can be traced to bird droppings and feathers entering a tank through an improperly designed and constructed manhole and improperly maintained vent screens.

Overflow. In order to keep insects, birds, and animals from entering the tank, the overflow screen, elastomeric check valve, and flap gate must be in place and must seal tightly. Overflow pipes should not extend directly into storm sewers or streams without an adequate air gap to prevent a possible cross-connection or backflow.

The overflow brackets should also be inspected to uncover any broken or cracked brackets or welds. Overflow pipes on tanks are intended for occasional use only. Tanks should not be regularly overflowed, and the overflow should not be used as a visual control for pumps and valves. Extreme overuse of the overflow pipe may damage the pipe or brackets. Trickling overflowing water can freeze and obstruct the overflow pipe.

Manholes. Roof manholes should be equipped with locks to prohibit unauthorized entry into the tank. Tanks that have access tubes leading to the roof should have their roof manholes properly latched to prevent them from blowing open in a strong wind, and any access doors to the tank ladders should be locked. Shell manholes should be properly sealed to prevent leakage.

Vents. Vent screens should also be in place and fit properly to prohibit the entry of insects and birds. Special vent designs may be necessary to prevent vents from clogging or freezing over. If an operator suspects that the tank vents have a tendency to become clogged or frozen over, the problem should be addressed by an engineer familiar with water tank vent design.

Wind or Earthquake Damage

If any tornado, major windstorm, or earthquake hits a tank, the tank should be professionally inspected to ensure that no damage occurred to the structure. In addition, operators should routinely look for possible wind or earthquake damage. Such damage on tower-supported tanks may be indicated by cracked coating or welds at the tower connections; broken, bent, or sagging rods; buckled struts; dented or twisted columns; or missing or loose rod pins. If any of these conditions is observed, the tank should be professionally inspected. In addition, it should be noted that tanks in areas at high risk for wind or earthquake damage should be inspected more frequently (every 1 to 3 years) than tanks in low-risk areas.

TANK WASHOUTS

As water is held in the tank, suspended solids settle out of the water into the tank bottom. Without regular washouts, tanks may accumulate large amounts of sediment. Sediment and deposits on tank walls decrease the effectiveness of disinfectant use. In addition, proper inspections cannot be conducted if sediment covers the bottom of the tank. Tanks should be washed out and inspected at least once every 3 years, and where water supplies have sediment problems, annual washouts are recommended. These washouts can be performed by the tank owner's personnel or a maintenance company.

Draining the Tank

Draining a tank may not be as easy as it first appears. Many times the valves are difficult to find; or once found, they may be inoperable. Therefore, before work crews are scheduled to wash out a tank, it is a good idea not only to map the valves in the distribution system but also to physically find and turn the necessary valves to block the tank off from the system and then drain the tank. Some tanks may be equipped with drain valves located in the wall or riser of the tank to facilitate tank drainage. These valves should be specially designed to prevent freezing.

Operating While the Tank Is Out of Service

Many cities, towns, and municipalities neglect to inspect their tanks because of a fear of operating without the tank. Proper operation while the tank is out of service usually requires that a relief valve be installed on a hydrant so that pressure fluctuations do not cause portions of the distribution system to become overpressurized. In addition, local businesses and individuals may need to be notified when a tank is going out of service so that temporary large uses of water (e.g., lawn watering or equipment washing) can be scheduled for other days, leaving the municipality with fire protection capacity. In some areas, portable pressure tanks may be available for limited storage.

Washing Out the Tank

The tank can be washed out with low-volume, moderate-pressure (2,400 psi [16.5 MPa]) pumps, firefighting equipment, or other means. The water should be sprayed on all surfaces to remove as much residue as possible. In areas where water pollution due to sedimentation is a problem or where strict local environmental regulations apply, it may be desirable to isolate the sediment and washout water and properly dispose of it, instead of allowing it to enter a storm sewer or nearby streams. In addition, care should be taken to prevent large amounts of sediment from entering the tank piping because pipes could be clogged, leading to valve damage.

If the tank is equipped with aluminum cathodic protection anodes, it is possible for many of these anodes to fall and remain in the tank. These anodes should also be removed at the time of the washout.

Disinfecting the Tank

The disinfection of water-storage facilities should be done in accordance with ANSI/ AWWA C652, Standard for Disinfection of Water-Storage Facilities. This standard describes three chlorination methods by which the disinfection can be accomplished.

This page intentionally blank.

Chapter 9

Professional Examination and Renovation

This chapter discusses professional examination and renovation of steel water-storage tanks. Maintaining steel water-storage facilities is increasingly important in view of rising replacement costs and the growing difficulty of obtaining rate increases or other funding for large-scale construction projects. Although maintenance costs are also increasing, small outlays for maintenance can substantially delay or even eliminate the need to replace a utility's large capital investment in tanks.

Residents and businesses located near tanks are concerned with the tank's appearance; and local, state or provincial, and federal agencies are now issuing new rules and regulations, policy statements, and directives that exert more stringent control over the maintenance and sanitary conditions of stored potable water. Of primary concern are the removal, transportation, and disposal of lead or chromate-based coatings.

The inspection and maintenance of the sanitary integrity of tanks has been largely ignored in the past. It was not mentioned or alluded to in AWWA D101. Water storage tanks can be a major source of contamination or recontamination in a water distribution system. With the increased concern for and regulation of drinking water, the maintenance of the sanitary integrity of water storage tanks demands special attention.

Maintenance needs apply equally to both a new steel water-storage tank and the rehabilitation of an existing tank. When renovation of an existing tank is being considered, the cost required to renovate the existing tank and its estimated remaining life must be evaluated; then the cost must be amortized and compared to that of a new tank to determine if the renovation is economically justifiable. A good, comprehensive preventive maintenance program can extend the life of an existing tank (as well as that of a new tank) indefinitely.

Many thousands of dollars can be saved and complaints from citizens eliminated if a planned approach to tank maintenance is adopted. This approach requires the expertise of experienced personnel to supervise the continuing inspection and maintenance of the utility's tanks. This individual may be one of the utility's own employees or someone employed by a professional inspection firm or other third-party firm. The role of the tank maintenance engineer is discussed in this chapter.

TANK MAINTENANCE ENGINEER'S FUNCTIONS AND QUALIFICATIONS

The general functions and qualifications of a tank maintenance engineer are the same, regardless of whether that person is a utility employee or a third-party consultant. In either case, qualified personnel are essential to a successful maintenance program.

Functions

A tank engineer/purchaser's representative is used during three distinct phases of the maintenance effort:

1. An engineer/purchaser's representative performs an evaluation or pre-bid inspection to indicate the condition of the tank and, if necessary, to serve as a basis for preparing maintenance specifications and bid documents.
2. Using the findings and recommendations of the engineer/purchaser's representative (from phase 1), an engineer prepares complete and definitive detailed technical specifications and bid documents. This identifies potential problems and reduces the need for change orders during construction.
3. An engineer/purchaser's representative performs competent full-time inspection of the work being done by the maintenance constructor to ensure strict compliance with the detailed technical specifications, drawings, and bid documents.

A maintenance engineering program will be most effective when all three phases are completed by the same individual or organization. If the three functions cannot be performed by the same firm, all parties must understand each other's responsibilities and keep communication lines open. A weakness in any one of the phases can reduce the effectiveness of the program and the quality of work for the money spent.

Qualifications

The tank maintenance engineer should have knowledge of the traditional engineering disciplines and have specialized training and practical experience in the design, fabrication, erection, inspection, sanitary integrity, coating, and maintenance of steel water-storage facilities. Engineers and purchaser's representatives must be experienced in the safe use of specialized rigging and other equipment to investigate the condition of the tank surfaces and to perform related functions required for tank maintenance. The owner should evaluate the purchaser representative's or engineer's qualifications in several areas.

Knowledge. The most important qualification for a purchaser's representative is up-to-date knowledge of the industry standards and requirements for work being done. The purchaser's representative must have a working knowledge of rigging; steel fitting and welding; surface-cleaning techniques; coating materials; coating application methods; cleaning and coating standards; and testing and inspection methods used to evaluate repair, cleaning, and coating operations. The purchaser's representatives must also be familiar with the relevant AWWA standards and standards of SSPC, ASTM International, and NACE International, as well as all regulatory agencies' rules and regulations and other recent industry standards. It is important for the engineer to understand the sanitary implications of the lack of screens and other openings in the top of tanks.

Effective communication skills. The purchaser's representative must be able to communicate effectively. These communication skills will be used to interpret and discuss the project specifications, as well as to resolve possible problems presented by concerned neighboring residents or businesses. The job of the purchaser's representative is to help the project run smoothly by noting problems and making constructive suggestions. A purchaser's representative who only points out deficiencies without facilitating solutions does not help achieve a timely completion of a quality project.

Climbing ability. The third main qualification is that the engineer is able to climb high structures. A purchaser's representative must closely monitor all parts of a constructor's work. The purchaser's representative must know proper rigging and safety practices and must maintain respect for the height at which he or she is working (Figure 9-1).

Equipment. At a minimum, the purchaser's representative should have and use the following equipment:

- camera
- daily inspection form
- SSPC written and visual standards for blast cleaning
- surface-profile measuring equipment
- wet-film gauge
- dry-film gauge
- certified calibration standards (for dry-film gauge)
- a minimum of two steel surface-temperature gauges
- psychrometer and psychrometric tables
- wet-sponge holiday detector
- Tooke Gage (for spot inspection)
- coating manufacturer data sheets
- flashlight
- weather radio
- safety equipment
- project specifications

Other items required may include equipment and instructions for dye-penetrant testing, vacuum-box testing, and radiographic or ultrasonic evaluation.

Figure 9-1 Experienced riggers evaluate hard-to-reach areas on tower tanks

PRE-BID INSPECTION

The pre-bid (initial) inspection is the first phase of the maintenance program. The purpose of this inspection is to gather data needed to estimate the work required and develop a budget for future expenditures. If the structure is in need of major renovation, the pre-bid report should provide information needed to make an informed decision on whether to maintain the tank or replace it. A secondary purpose might be to define the condition of the tank if the owner is contemplating relocating it or selling it. The inspection report should inform the owner what degree of rehabilitation is necessary. The purchaser's representative should gather all information necessary to prepare drawings and specifications for recoating and repairs in the event that a decision is made to keep the tank in service.

The findings of the pre-bid inspection report should be presented in such a manner that the condition of the tank is evident regardless of the reader's background. One of the most important functions of this report is to convince those in charge of allocating funds that recommended repairs and maintenance expenses are necessary. Many small utility owners have a limited knowledge of tanks. Color photographs keyed to the observations are an excellent method of describing the condition of a particular portion of the tank.

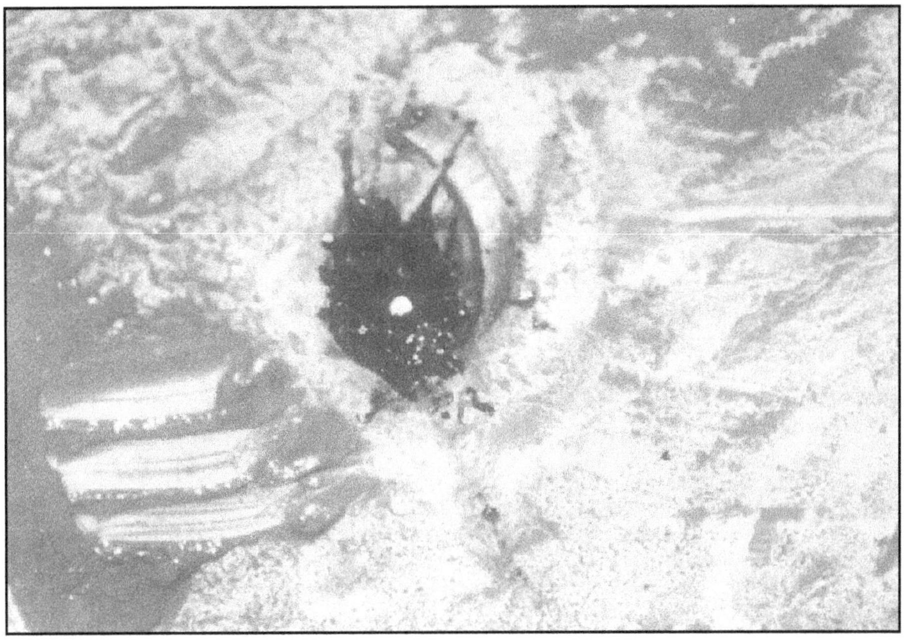

Figure 9-2 Active corrosion penetrated this ¼-in. (6-mm) steel tank bottom in 9 years. Periodic inspections and washouts would have revealed and prevented this problem well in advance of failure.

Components of the Inspection

The tank should be evaluated from several standpoints to provide an accurate evaluation of its condition. Individuals entering the tank should comply with Occupational Safety and Health Administration (OSHA) or state or other applicable regulations concerning entry into confined spaces.

Foundation. Sufficient information should be gathered to determine the structural integrity of the foundation. The foundation should be checked for visible concrete deterioration as well as hollow-sounding areas. The purchaser's representative can find concrete laminations by tapping the concrete with a hammer. The concrete foundation of a tank tends to show a greater level of deterioration than any other structural component. In particular, evidence of foundation settling should be noted and its cause evaluated.

Supporting tower. The diagonal and riser rods should be inspected for any rust or metal loss that may reduce their strength. They should also be checked for proper tension. All of the connections (clevises, wing plates, turnbuckles, and strut-to-column connections) should be checked for structural integrity. The bolts and welds at these connections should be inspected for metal loss and deterioration.

Container. The container of the tank should be inspected on the interior and exterior to determine the extent of metal loss and the remaining membrane thickness (Figure 9-2). Rivets and bolts should be examined for tightness and metal loss. Pits should be measured with a pit-depth gauge to determine the extent of the pitting. The steel around the pit should be measured with an ultrasonic thickness gauge (Figure 9-3) to determine the thickness of the steel surrounding the pit. The remaining steel thickness at the bottom of the pit can be calculated by subtracting the pit depth from the ultrasonic thickness measurements. With this information a resultant net membrane thickness can then be calculated to determine the stress

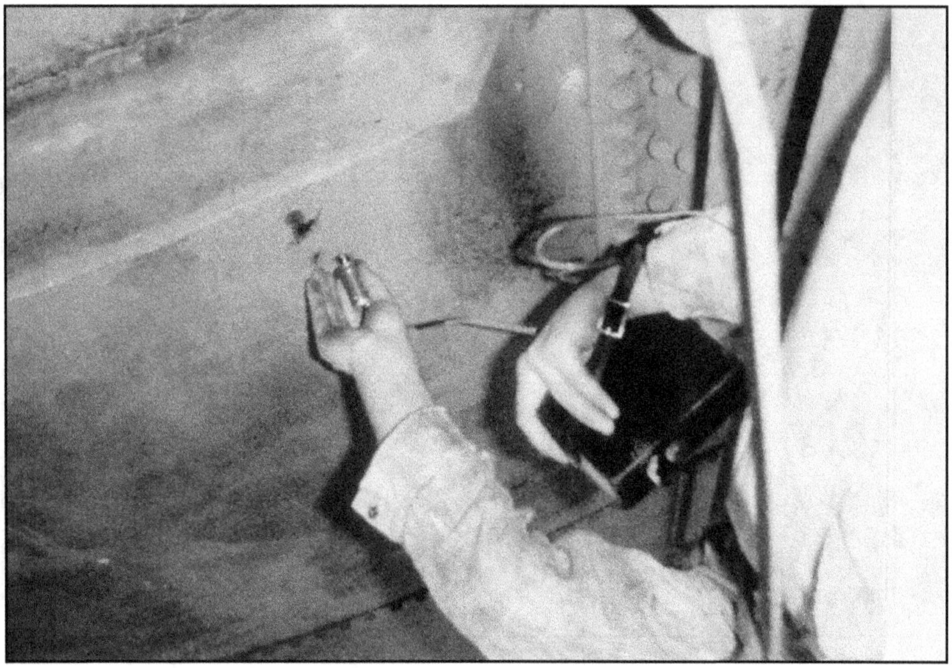

Figure 9-3 Measuring shell thickness with ultrasonic equipment

levels in the remaining steel. When the steel has experienced general metal loss across a large area, multiple ultrasonic thickness measurements should be taken to determine the membrane thickness.

Coating. The condition of the coating on the tank should be evaluated. To accurately determine a coating or recoating schedule, it is necessary to estimate the remaining effective life of the coating system. The owner should define the criteria for effective life based on both aesthetics and corrosion protection. Factors to consider might include amount of rust, peeling, chalking, or graffiti. Once the criteria have been defined, one can project the coating life by monitoring the deterioration of the coating based on the selected criteria. The thickness, type, and adhesion of old coatings should be evaluated prior to preparing specifications for recoating. Adhesion tests, dry-film thickness, and estimation of percent coating failure to rust and primer all are factored into estimating the effective remaining life. Tests should also be performed to identify the generic type of coating presently on the tank, in order to determine the possibility of topcoating and the potential need to dispose of any removed coatings as hazardous waste. For example, lead-based coatings will quite likely be considered hazardous waste.

Sanitary conditions. The sanitary integrity of various components of water tanks have come under increasing scrutiny in the past several years. The tank piping (including drain and overflow) should be checked for possible cross-connections. All of the manholes, hatches, vents, and openings should be checked for compliance with AWWA standards and local, state or provincial, and federal regulations. One of the environmental regulations drawing public attention lately deals with coatings that contain lead and chromates. These coatings are considered a potential health threat if they are cleaned from the tank and allowed to escape into the atmosphere or onto the ground. The container's interior should be washed out on a regular basis to clean it of debris and make it easier to inspect (Figure 9-4).

Figure 9-4 Washing out tanks allows easier inspection and keeps tanks sanitary

Vent and overflow. Look for missing, damaged, rusted, or warped screens and check that screen sizes can prevent birds and insects from entering the tank. Check that screens are properly located so that debris on the screen will not drop into the tank. If frost is a possibility, ensure that there is provision for pressure and vacuum release in the vent or by means of an auxiliary relief device.

Also check the vent and overflow for proper size if there has been a significant increase in water usage and a corresponding increase in the fill/withdraw rates of the tank.

Access hatches. Ensure that they are properly installed, fit snug, can be easily secured, and have a 4-in. (100-mm) curb or neck and 2-in. (50-mm) downward overlap, or are tightly gasketed, bolted, and sealed.

Finial balls. Check that they are effectively screened. Finial balls are notorious for allowing birds and insects into the tank and are difficult to screen. In most cases, they should be replaced with a more acceptable vent.

Holes in the roof. Check for holes cut by coating companies or cathodic protection installers or formed by rust or popped rivets and of any other openings that may allow rainwater to drain into the tank.

Roof overhang. Check for gaps between the roof and vertical wall. This is normally not a sealed connection in older tanks, and warping from uneven expansion and contractions will result in large gaps.

Cathodic protection. A cathodic protection system, if installed, should be evaluated and properly maintained. Systems with manual controls and temporary anodes require yearly maintenance to adjust the potential and replace deteriorated anodes. Newer impressed-current systems with automatic controls and long-life anodes require less maintenance and are therefore more easily forgotten. However, these systems should also be checked periodically during the year by water system personnel to make sure that they are operating, and they should be checked every year by a corrosion specialist for proper operation.

Appurtenances. The tank accessories and appurtenances should be inspected to ensure proper operating conditions. These items include level gauges, access lighting, obstruction lighting, and antennas. The ladders, safety railings, platforms, and other safety-related items should be inspected for deterioration and to ensure that they meet all applicable regulations.

Environment. The tank inspection should include an evaluation of the tank's environment and operating conditions. Factors such as the tank's location with respect to buildings, parking lots, people, and other structures that would affect the exterior coating life and tank maintenance operations should be noted. Operating characteristics of the tank, such as susceptibility to freezing, sweating (condensation on the exterior of the tank), and the aggressiveness of the water stored, should be noted and referred to when specifications are being prepared.

Evaluating the Inspection Information

Before the report data are assimilated, some evaluations and calculations should be made. The information regarding interior pitting should be carefully evaluated as to location, depth, and density of metal loss. This evaluation requires specialized experience with steel tanks. Thousands of random pits on the interior surface of a tank may not be detrimental to the structural integrity of the container; however, a small number of long vertical-groove pits in a specific area can significantly reduce the membrane strength of the container shell. Based on the information from the inspection, structural engineers can calculate the adequacy of the remaining membrane strength of the steel. Pitting in bolted tanks may be more critical than in welded tanks because of the difference in minimum plate (sheet) thickness and allowable design stresses.

If warranted by the information gathered from the inspection, the adequacy of deteriorated members and connections in the support structure should also be calculated.

The engineer/purchaser's representative performing the tank evaluation must be competent in designing tanks. This qualification is important because someone not competent in this area will not be able to interpret the inspection results accurately. The results of the tank evaluation or pre-bid inspection should be certified by a professional engineer. Some states do not recognize engineers as being licensed unless they are registered in the state where the project is taking place.

Recommendations

After the evaluation is complete and the observations listed, a list of recommendations should be compiled based on the purchaser representative's observations and the engineer's calculations. The recommendations, alternatives, and cost estimates will help the decision makers arrive at an informed decision.

Structural repairs. The recommendations should note all structural modifications that the engineer determines are necessary. These recommendations may range from minor welding to replacing large sections of steel plate. If the evaluation is performed properly by a registered professional engineer experienced in tank design and maintenance, the repair procedures can be more exactly defined, potentially saving a significant amount of money. For example, expert evaluation can minimize the number of pits that need to be welded or identify a structural member that can be adequately stiffened without replacement.

Coating systems. The recommendations should give options for coating systems, with a complete description of the advantages and disadvantages of each. These advantages and disadvantages should be based on the characteristics of the coating type in conjunction with the tank environment and operating conditions.

Bolted tanks. For bolted tanks coated with fused glass, an entire steel sheet can be removed and replaced if the glass coating is severely damaged. Leaking seams or small breaks in the glass coating can be repaired with the polyurethane sealant used to seal the seams initially. Galvanized tanks can be coated with a zinc-based or epoxy coating to cover damaged spots in the galvanizing. Epoxy- and polyurethane-coated bolted tanks may be abrasive blast-cleaned and recoated. Since the thin steel on bolted tanks usually precludes repair by welding, penetrating pits are usually repaired by a bolt placed through the hole in the tank. However, if large areas of the coating have failed, it may be more economical to replace tank panels.

Sanitary or safety hazards. The recommendations should also include modifications and additions to all portions of the tank that present a sanitary or safety hazard. AWWA, OSHA, ANSI, and other federal, state or provincial, and local regulations and standards should be checked to ensure that all sanitary and safety codes are met and that birds, insects, and animals are excluded from the tank. Recommendations for enlarging the vent or overflow should be made if they are too small because of a significant increase in the size or capacity of the water system.

PREPARING SPECIFICATIONS

Once a complete evaluation of the condition of all of the tank components has been made and it has been determined that repair of the tank is required and economically feasible, a set of detailed technical specifications, drawings, bidding, contracting, and bonding requirements can be prepared.

Determining the Scope of Work

The scope of work should be determined by evaluating the recommendations and cost estimates from the inspection report and comparing these to the available funds. It is usually not a good idea to coat only the exterior or interior of the tank if both areas need work. In most locations, the tank must be drained for either exterior or interior coating. (Condensation will occur on the tank exterior if the water and steel temperature are below the dew point of the air.) Splitting the work also incurs the cost of a second mobilization by the constructor.

Writing the Specifications

Once the scope of work has been defined, the specifications must be written. To write the specifications, the specification writer should have a complete knowledge of the tank's condition and environment. Thorough reading of the inspection report should give the writer all the necessary information.

Coating types. The specification writer should be familiar with the types, availability, and applicability of modern coating systems for both the interior and exterior of the tank. The environmental characteristics of the tank site should be considered when the type of coating to be used is selected. Some possible items of concern are the proximity of buildings and cars, corrosive fumes discharged by factories, and salt-air environments. Each of these situations presents unique problems for the coating applicator and for the performance of the selected system. The coating system selected should minimize problems for the applicator and give a satisfactory combination of life and cost under the present and anticipated environmental conditions.

Site restrictions. The specification writer should be aware of any site restrictions that need to be addressed. Examples of some items that might be included in the specifications are

- covering sensitive equipment in the area to protect it from abrasives and coating
- limiting working days because wind conditions may blow coating spray into the nearby water treatment basins or onto adjacent property
- covering insulation on piping to prevent damage
- limiting working hours to certain times of day because of a hospital in the area

Industry standards. The specification writer needs a working knowledge of, and access to, ANSI/AWWA D100, Standard for Welded Carbon Steel Tanks for Water Storage; ANSI/AWWA D102, Standard for Coating Steel Water-Storage Tanks; ANSI/AWWA D103, Standard for Factory-Coated Bolted Carbon Steel Tanks for Water Storage; ANSI/AWWA D104, Standard for Automatically Controlled, Impressed-Current Cathodic Protection for the Interior Submerged Surfaces of Steel Water Storage Tanks; ANSI/AWWA D106, Standard for Sacrificial Anode Cathodic Protection Systems for the Interior Submerged Surfaces of Steel Water Storage Tanks; ANSI/AWWA D107, Standard for Composite Elevated Tanks for Water Storage; ANSI/AWWA D108, Standard for Aluminum Dome Roofs for Water Storage Facilities; ANSI/AWWA C652, Standard for Disinfection of Water-Storage Facilities; the Society for Protective Coatings (SSPC) *Systems and Specifications*; OSHA/ANSI regulations; and the rules and regulations of all other appropriate regulatory agencies. The writer should also be familiar with the capabilities and availability of constructors qualified to perform work of the nature and magnitude required.

Completeness. One of the most important functions of the specifications is to address potential problems, loose ends, or gray areas before the start of the project. This will ensure that the work will proceed as smoothly as possible and that the finished product will be of high quality. Although it is difficult to determine the condition of every square inch of the tank before the steel is blast cleaned, steps can be taken to reduce the impact of change orders and make them more equitable to the owner and constructor. One of the best methods is to have the constructor include unit prices for a predetermined quantity of different repairs along with the bid. The unit price should be a readily measurable unit that the owner and constructor can reasonably determine. The specifications should also state that unfair pricing or unbalanced unit prices are causes for rejecting a bid. Methods should be established to resolve disputes among the owner, constructor, and neighboring property owners.

Start and finish dates. Guidelines should be included for anticipated start and finish dates. Factors to consider in determining the required time frame for the work include allowances for system operating requirements, anticipated weather conditions during the time of work, application characteristics of the selected materials, workloads of the prospective bidders, time of year bids are taken, and unusual site conditions. It would be unreasonable, for example, to require a constructor to complete a project during the spring rainy season with no weather allowance, or to specify a two-component epoxy coating system on the interior and not allow at least 1 week (at 70°F [21°C]) for curing. Tank out-of-service times depend on the size of tank, the scope of work, the weather, and the coating system.

The specification should also address the sequence of operations. For example, all cutting and welding should be performed on both sides of the steel before any cleaning and coating are done.

Bonding and insurance. Bonding, insurance, and legal documentation must be incorporated into the contract documents. The constructor should be required to have performance and payment bonds in the amount of 100 percent of the contract. Bid bonds may also be required. Most utilities, governmental units, or municipalities have unique legal requirements that must be incorporated into the specifications. A first-anniversary inspection is recommended and should be performed in accordance with AWWA standards. The inspection should be conducted before expiration of the 1-year bonded guarantee that is provided for in the performance bond.

Constructor investigation. If possible, the bidders should be thoroughly investigated by the owner or engineer. The extent of the investigation will depend on the owner's or engineer's knowledge of the constructor. The constructor's capabilities, experience, present schedule and workload, and financial strength should be evaluated. It may also be advisable to inspect some of the bidder's previous work and to investigate the firm's financial position.

MONITORING THE CONSTRUCTOR'S PROGRESS

Once the contract has been awarded, the owner must be assured that the constructor will follow the specifications as written. The only way to ensure full compliance with the specifications is through quality control inspections performed by the owner.

The key to the longevity of any tank repair and coating project is the quality of the work. The purchaser's representative should substantiate any requests for contract change orders submitted by the constructor and the quantities of work done on a unit-price basis. Without a purchaser's representative on the job, there is no way to confirm this extra work. The two types of quality control inspections are full-time and spot.

Full-Time Resident Inspection

Under full-time resident inspection, a qualified, experienced inspector hired by the purchaser is on the site at all times while work is being done.

Tank repairs. The purchaser representative's first responsibility in regard to the repair work is to verify that the welders and other workers are fully qualified in their respective fields. All fabricated appurtenances should be checked for compliance with the specifications and drawings before installation. The purchaser's representative should witness or verify that all required testing of the repair work has actually been accomplished. Finally, the purchaser's representative must be able to confirm that all repair work (e.g., orientation, installation) has been done in accordance with the technical specifications and drawings.

Visual standards for cleaning. The purchaser representative's first responsibility during the cleaning and coating phase of the project is to set forth the guidelines for acceptable blast cleanliness to which the constructor must comply. NACE and SSPC Vis 1 are visual standards for steel cleanliness that may be used by the purchaser's representative as a basis for the guidelines (Figure 9-5). An area of acceptable cleanliness can be marked off and coated with a clear polyurethane coating for reference during the project. Additionally, before coatings are applied, the surface should be checked for the presence of chlorides, sulfides, or other chemical contaminants. An abrasive blast-cleaning operation is shown in Figure 9-6.

Surface profile. After steel cleanliness is determined, the purchaser's representative must continually monitor the surface profile of the steel to be sure it falls within the coating manufacturer's recommendations. The surface profile is the root-mean-square distance between the peaks and valleys in the steel surface. Surface roughness forms the mechanical bond between the coating and the steel; however, if the profile is

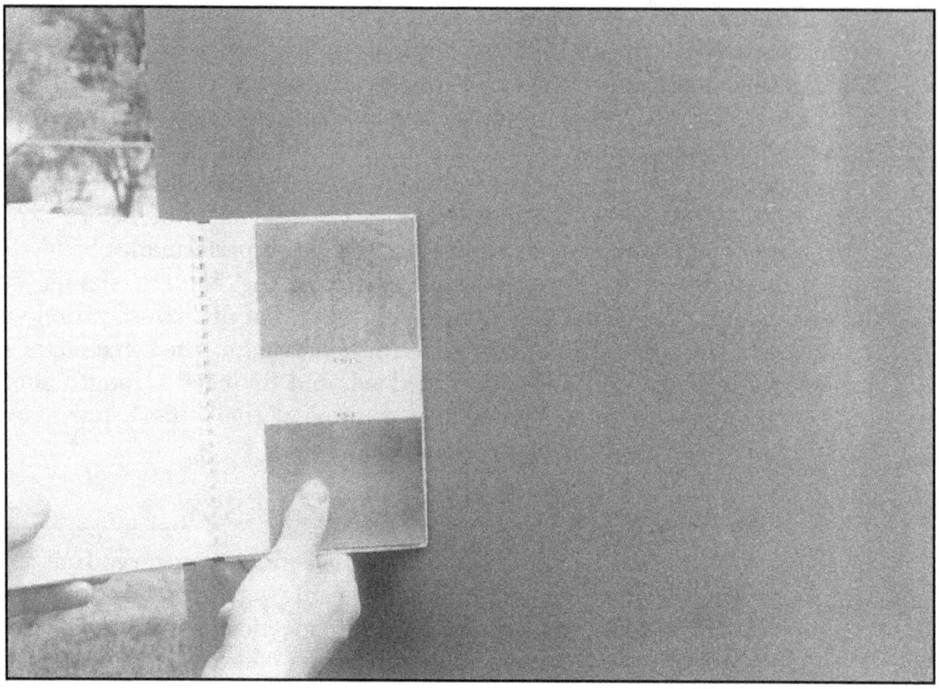

Figure 9-5 Inspection of the degree of abrasive blast cleaning

Figure 9-6 An abrasive blast-cleaning operation

too deep, the coating system may not be able to cover all of the peaks. Surface profile can usually be controlled through the choice of blasting grit size.

Coating application. After the steel has been properly cleaned, the purchaser's representative should monitor the coating application process. Specifications are usually written in terms of the dry-film thickness of the coating system. For most coatings the wet-film thickness is different from the dry-film thickness; however, multiplying the wet-film thickness by the percent volume solids of the particular coating being applied provides a good approximation of what the dry-film thickness will be after curing. Measuring the wet-film thickness gives only a guideline as to whether or not the constructor is applying the coating at its proper thickness. The dry-film thickness must still be measured to determine if the coating system thickness meets the specifications.

Atmospheric conditions. The atmospheric conditions should be continually monitored and recorded to ensure that the manufacturer's recommended limits for humidity and temperature are being followed. The dew point should be determined from dew-point charts. When the steel temperature falls to within 5°F (2.8°C) of the dew point, all coating operations should be suspended. This 5°F (2.8°C) buffer is a safety factor to ensure that moisture does not condense on the surface of the steel or uncured coating. It also allows for the inaccuracies of the measuring equipment. It is important to monitor conditions on the tank site instead of relying on National Weather Service data. Conditions can vary widely only a few miles from the reporting station. Conditions at an airport, which is surrounded with concrete, are often not the same as those at the tank site in a grass-covered field.

Coating batch numbers. All batch numbers from the cans of coating used on the project should be recorded and referenced to the portion of the tank on which the can of coating was applied. If the coating requires that two or more parts be mixed, the mixing should also be monitored by the purchaser's representative. These records will allow the cause of any coating failure to be more easily identified.

Thinners. The use of thinners should be only as prescribed by the coating manufacturer. Not only can improper use of thinners affect the serviceability of a coating, but also additives may render the coating unfit for potable water contact.

Holiday testing. A wet-sponge holiday test should be performed on all water-bearing surfaces at the conclusion of the project. (Thicknesses in excess of 20 mils usually require high-voltage spark testing for holidays.) The holiday test detects small voids and discontinuities in the coating system, which may be undetectable to the naked eye. If the holiday detector responds, it means that water from the sponge is reaching the steel substrate and completing the electrical circuit. If water is able to reach the steel, it will not be long before the coating fails at the defect. The holiday should be touched up to prevent premature failure.

Spot Inspection

Spot or intermittent inspections encompass the same procedures and philosophies as a full-time resident inspection. However, with this type of inspection, the owner provides a capable employee to perform resident daily inspections, and a professional firm periodically monitors the progress.

The frequency of professional inspections depends on the reliability of the crew doing the work and the expertise of the owner. At a minimum, tanks should be inspected by a professional at the following intervals:

- at the beginning of blast-cleaning operations, to establish an understanding and agreement between the constructor and purchaser's representatives regarding the degree of cleaning required on the tank surfaces

- at the start of the application of the primer coat
- before the finish coat is applied
- once work is completed, to verify final coating thicknesses, acceptability of application, continuity of interior coatings, and acceptability of the exterior coatings' appearance.

Because it is difficult to know exactly what took place between inspections, the professional performing spot inspections may have to do a considerable amount of research. Full-time inspection of tank coating is recommended because of (1) the technical nature of tank coatings, (2) the need to comply with environmental regulations, and (3) the need to apply coatings intended for potable water contact in such a manner as to maintain the product's certification according to NSF/ANSI 61.

Unit-price work and work being done by change order should be performed in the presence of the purchaser to ensure the work is being done in the owner's best interest.

PERIODIC REINSPECTION

Once the three phases of tank maintenance have been completed and the first-anniversary inspection has been performed, the tank should be inspected at least once every 3 to 5 years or as required by state regulatory agencies.

Three methods of evaluating tank interiors are currently popular: a drained "dry" evaluation, an underwater evaluation performed by divers, and a robotic inspection.

The dry evaluation is the most thorough evaluation, as the tank can be washed out and the mud and debris removed from the interior surfaces before the inspection. Specially trained technicians can get to the necessary surfaces on the tank by rigging and rappelling the interior as required by the condition of each tank.

When the tank cannot be drained, an underwater evaluation of the tank interior may be performed. The diving evaluation should include a careful study of the tank's interior wet surfaces by a Certified Commercial Diver who is experienced in the inspection of water tanks. When performing an underwater evaluation, the diver should use a full dry suit and full-face diving mask to prevent contamination of the potable water. Prior to entering the tank, the diver and all equipment are thoroughly disinfected in accordance with ANSI/AWWA C652-11. The tank must be isolated from the water system and all connecting valves locked out.

The third type of evaluation is a remotely operated vehicle (ROV) inspection of the tank interior. This type of evaluation utilizes a robotic device, guided and monitored by a specially trained field technician, to videotape the tank interior. The video should then be reviewed by a professional engineer experienced in water tank evaluation to determine the condition of the interior surfaces.

These inspections will turn up minor deficiencies that can be corrected before they become major, expensive, and time-consuming problems. Cathodic protection may be installed to protect the steel from metal loss in case small coating failures develop. Repairs and cathodic protection can significantly extend the life of a coating system. With periodic inspection, cleaning, and touchup work, a steel tank can remain in service for many decades without major renovation.

Chapter 10

Cold-Weather Operation

A frozen water-storage tank can cause more problems than no tank at all. Fortunately, proper specification of appurtenances and operating procedures will eliminate most tank freezing problems. This chapter discusses ways to prevent cold-weather problems associated with elevated tanks, standpipes, and ground storage reservoirs. The chapter also gives suggestions on dealing with these problems if they do occur. Figure 10-1 shows an extreme example of a frozen tank.

CAUSES AND RESULTS OF FREEZING

In general, water freezes when the atmospheric conditions cause heat to escape faster from the water than it can be replaced. The following material discusses conditions that can cause freezing and problems that can result.

Causes of Freezing

Three general categories of problems may cause freezing and, subsequently, ice damage to water-storage tanks. The first category is improper specification. The second and third categories, static water conditions and overflow, are essentially operational problems.

Improper specification. The first general cause of freezing is improper specification of the tank or the control systems. Specification elements that can influence freezing problems include

- piping arrangement and size
- vent design
- inadequate amount of freeboard between the overflow and the roof structure
- structural members, piping, ladders, or other protrusions below the high water level
- heat-conducting projections or narrow gratings across riser piping

Figure 10-1 A frozen water tank

- level of insulation on the piping or riser
- piping projections from the tank in which the water is static

Static water conditions. Static water conditions, or lack of sufficient water turnover, prevent water in the tank from being warmed by incoming water. Some of the causes of this condition are as follows:

- Some tanks are reserved for fire protection and so are not used regularly. (These should be heated if the climate dictates.)
- The tank top capacity level may be lower than the system pressure gradient and the water may be kept static by an altitude valve.
- The tank may be located in a low-water consumption area, with high pipe friction between the tank and the pumping facility and between the tank and the high-use area.
- A normally heavy water user may suddenly stop using water (such as during a school vacation or when a factory is shut down for several days).
- Pumps supplying the system may be of a variable-output, constant-pressure type and may have a capacity equal to the peak demand.

- The water main serving the tank may freeze solid.
- The tank may be valved off for winter with water in it.
- The controls for the pumps or altitude valve may become frozen or damaged by ice.
- The tank may be too large for the present system.

Tank overflows. Freezing problems may also occur when the tank is allowed to overflow at a trickling or slow rate that permits freezing in the overflow pipe. In freezing weather, the escaping water will immediately freeze. Causes of overflow include:

- improper control specification or adjustment
- frozen controls for the pump or altitude valve
- telemetering conductors damaged by ice or wind
- leaky valves
- tanks designed to float on a system with other tanks (these will occasionally overflow because of pressure variations)

Results of Freezing

When a tank freezes, one or more of the following conditions usually results and causes leakage:

- Inside overflow or other piping breaks occur.
- Ladders or other attachments to the container are pulled out by ice, making a hole at the point of attachment.
- Ice pressure can expand and place hoop stress loads on the steel and seams, which may cause the tank wall to yield or burst.
- Leaks due to corrosion become apparent.

Overflow through overflow pipes. When tanks overflow in freezing weather, several problems can develop. An overflow-to-grade may freeze solid, especially where there is a trickling overflow or where screens are plugged or flap valves are stuck on the discharge. If water continues to be pumped into the tank after the overflow pipe is frozen solid, the tank may overflow for a while through the roof hatch, then through the vent; the tank will then freeze solid, build up pressure, and burst.

Overflow through the roof hatch or vent. When water overflows through the hatch and vent, it invariably forms a large icicle, weighing tons, on the tank exterior. The same problem can be caused by normal overflow through a stub overflow (one extending only a few feet from the shell of the tank). This places a large eccentric load on the structure, that, in the case of water towers, can exceed the structure's design stress. The icicle usually forms on the side of the tower that is away from the prevailing wind, and the wind and icicle together create additive loadings. Even if the structure is not damaged by the ice load, it may be damaged when the ice thaws or breaks off and falls. Eccentric ice loadings or tower members damaged by falling ice have caused water towers to fall.

Frosting over tank vents. A unique freezing problem may occur when frost freezes solid over the fine screen in the vent and overflow. This type of freezing will usually occur on the fine screen designed to keep insects out of the tank. Such freezing prevents the exchange of air into the tank, resulting in a vacuum in the tank that can collapse (implode) the tank until there is a structural rupture to break the vacuum.

DATA RELATED TO FREEZING

Weather data and heat-loss data are both significant in relation to tank freezing.

Weather Data

Lowest one-day mean temperatures (LODMTs) for the United States are charted on isothermal lines in Figure 10-2. Of course, weather conditions are very erratic, and a given location may experience temperatures lower than those exhibited on the LODMT chart.

Experience has shown that tanks located in areas with the LODMT warmer than –5°F (–15°C) normally will not experience cold-weather operational problems, with the following exceptions:

- elevated tanks with small-diameter (less than 36 in. [0.9 m]) uninsulated risers
- tanks with uninsulated and unheated appendages that experience no flow (for example, a drain valve at the end of a nipple)
- tanks experiencing overflow problems because of frozen or otherwise malfunctioning controls for valves or pumps
- tanks with abnormally static water conditions
- tanks with inadequate vents or overflows

Heat-Loss Data

The National Fire Protection Association publishes tables giving the heat loss per hour from various types of tanks. Table 10-1 presents this data for small elevated tanks. For example, a 250,000-gal (0.95-ML) elevated tank located in an area having a –10°F (–23°C) LODMT loses 1 million Btu/h (293 kJ/s) in a 12-mph (19-km/h) wind. In a –40°F (–40°C) location, the heat loss almost doubles.

DESIGNING TANKS FOR COLD WEATHER

Proper design of a tank will prevent most freezing problems and will allow personnel to follow operating procedures to deal easily with any that do occur.

Inside Appurtenances

Tanks located in an area where the LODMT is –5°F (–15°C) or colder should not be equipped with inside ladders or overflow pipes. As ice forms and moves up and down, it can exert tons of force on ladders and pipes, tearing them loose from their supports and possibly ripping or punching holes in the container. The resulting leak will occur at a very inopportune time. If an inside overflow pipe is broken, the tank will rapidly lose all water down to the break, creating a large icy area on the ground below. If the vent is plugged with ice or snow, the tank roof may collapse when water evacuates the tank rapidly.

Table 10-1 Thousands of British thermal units (Btu) lost per hour* from elevated steel tanks based on minimum water temperature of 42°F (5°C) and a wind velocity of 12 mph (5 m/sec).

Atmospheric Temperature °F	Heat Loss (Btu) per ft², Tank Radiating Surface	Tank Capacities, thousands of US gal									Additional Btu loss per lineal ft, Uninsulated Steel Riser	
		25	30	40	50	75	100	150	200	250		
		Square feet (ft²) of tank surface†										
		1,210	1,325	1,550	1,800	2,370	2,845	3,705	4,470	5,240		
		Btu loss per hour, thousands									3 ft dia.	4 ft dia.
35	32.3	40	43	51	59	77	92	120	145	168	50	69
30	46.1	56	62	72	83	110	132	171	207	242	144	192
25	61.5	75	82	96	111	146	175	223	275	323	255	340
20	77.2	94	103	120	139	183	220	287	346	405	380	506
15	93.6	114	125	146	169	222	267	347	419	491	519	692
10	110.9	135	147	172	200	263	316	411	496	582	670	893
5	128.9	156	171	200	233	306	367	478	577	676	820	1,092
0	148.5	180	197	231	268	352	423	551	664	779	982	1,309
–5	168.7	205	224	262	304	400	480	626	755	884	1,152	1,536
–10	190.7	231	253	296	344	452	543	707	853	1,000	1,329	1,771
–15	213.2	258	283	331	384	506	607	790	954	1,118	1,515	2,020
–20	236.8	287	314	368	427	562	674	878	1,059	1,241	1,718	2,291
–25	262.3	318	348	407	473	622	747	972	1,173	1,375	1,926	2,563
–30	288.1	349	382	447	519	683	820	1,068	1,288	1,510	2,145	2,860
–35	316.0	383	419	490	569	749	900	1,171	1,413	1,656	2,381	3,174
–40	344.0	417	456	534	620	816	979	1,275	1,538	1,805	2,620	3,494
–50	405.6	491	538	629	731	962	1,154	1,503	1,814	2,126	3,139	4,186
–60	470.8	570	624	730	848	1,116	1,340	1,745	2,105	2,467	3,702	4,936

*To determine heat loss per hour, find the minimum mean atmospheric temperature for one day from the isothermal map, Figure 10-2, and note the corresponding heat loss above.

†These numbers are square feet of tank radiating surfaces used for each capacity to compute the tabulated heat loss values and are typical for tanks with D/4 ellipsoidal roofs and bottoms.

NOTE: Heat loss for a given capacity with a different tank radiating surface than shown above shall be obtained by multiplying the radiating surface by the tabulated heat loss per square foot for the atmospheric temperature involved. The minimum radiation surface area shall be the wetted tank steel surface area plus the top water surface area. For tanks with large steel plate risers, the heat loss from the riser shall be added to that from the tank. The riser loss per linear foot shall be as tabulated above.

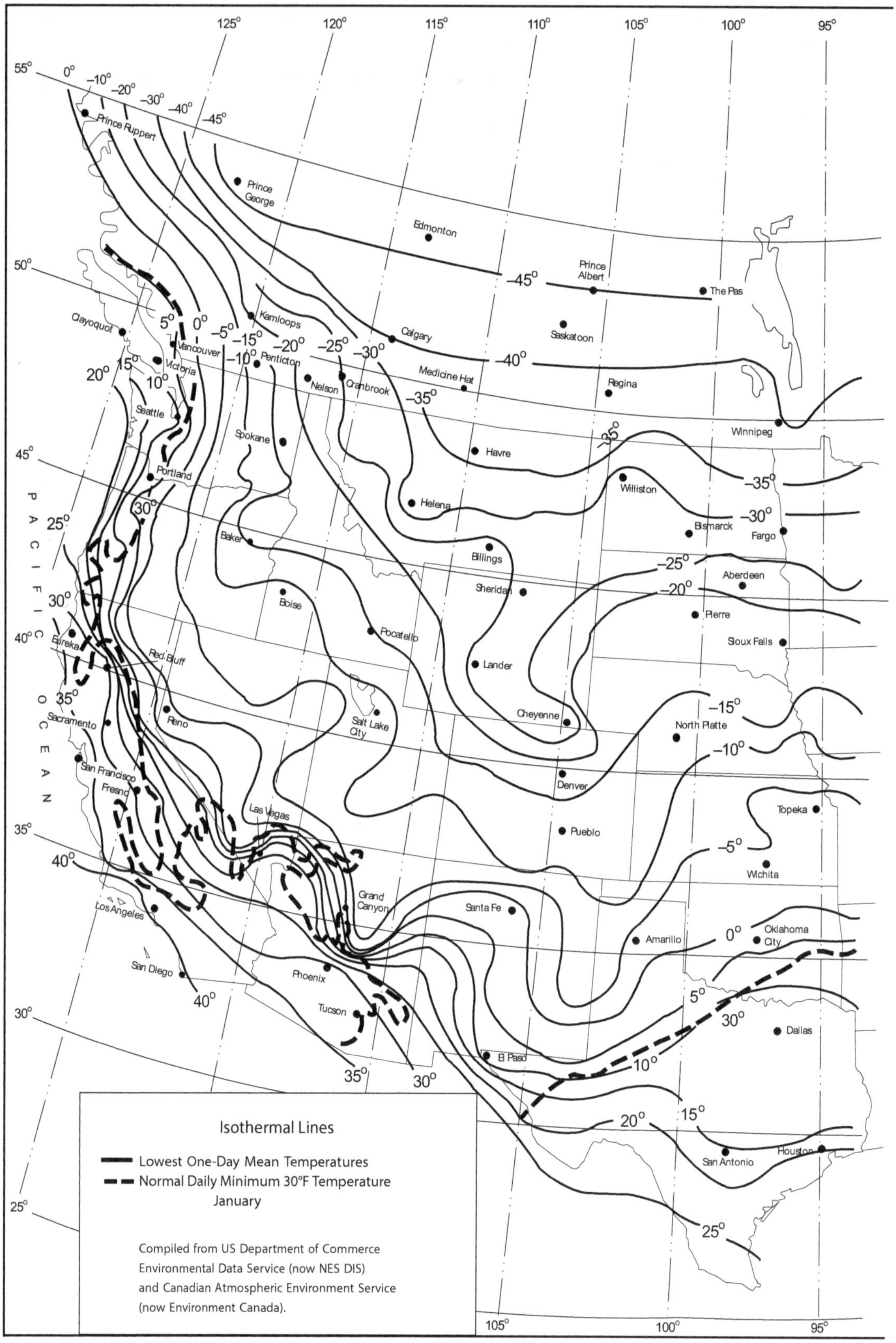

Figure 10-2 Isothermal lines for lowest one-day mean temperatures and normal daily minimum 30°F (–1°C) temperature line for January, United States and Southern Canada

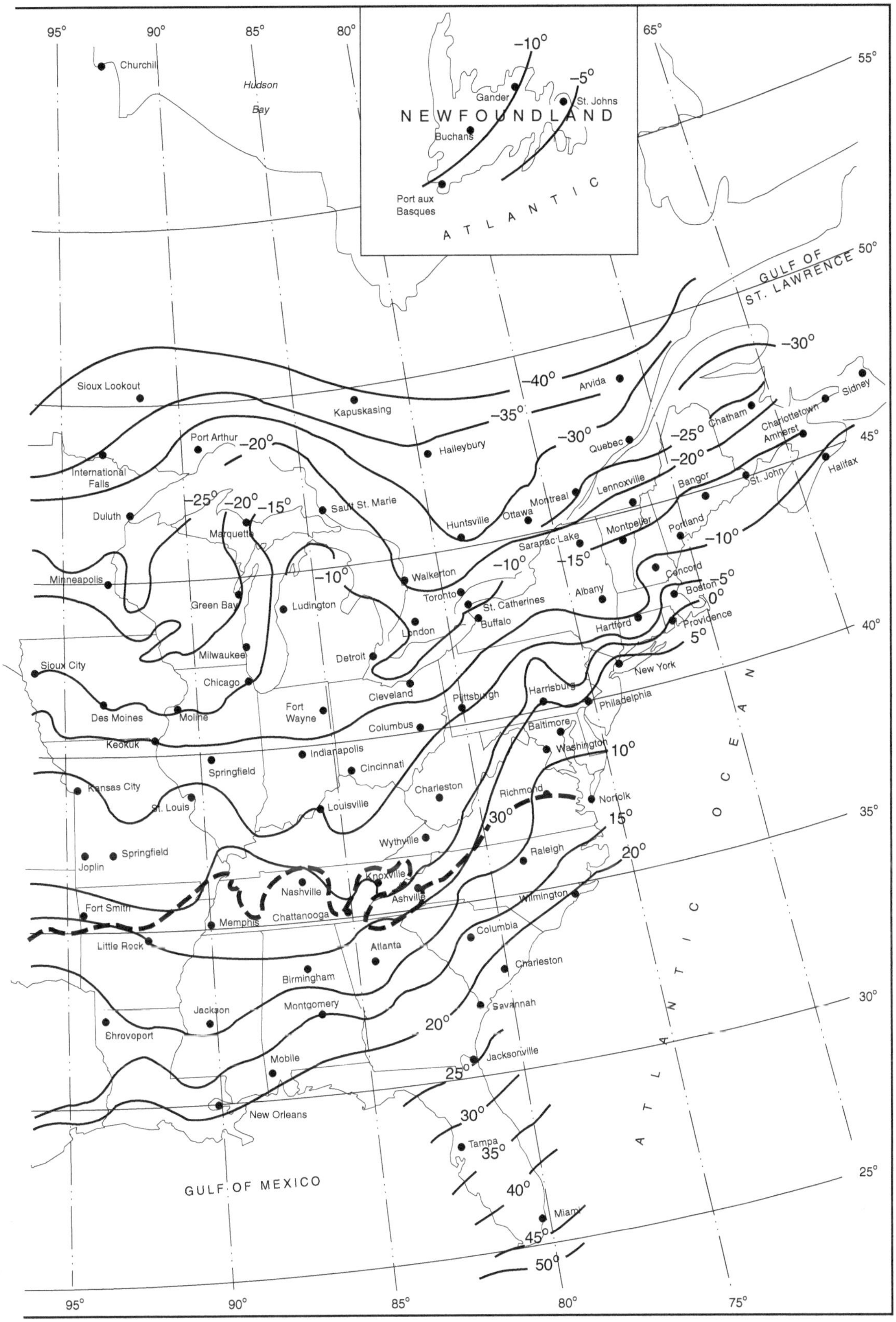
NEWFOUNDLAND
Gander
St. Johns
Buchans
Port aux Basques
ATLANTIC
−10°
−5°
Churchill
Hudson Bay
GULF OF ST. LAWRENCE
Sioux Lookout
Kapuskasing
Arvida
Sidney
Chatham
Charlottetown
Amherst
Halifax
Port Arthur
Haileybury
Quebec
St. John
Bangor
International Falls
Lennoxville
Montreal
Ottawa
Huntsville
Sault St. Marie
Duluth
Marquette
Saranac Lake
Montpelier
Portland
Concord
Boston
Providence
Minneapolis
Walkerton
Toronto
St. Catherines
Buffalo
Albany
Hartford
Green Bay
Ludington
London
Sioux City
Milwaukee
Detroit
New York
Chicago
Cleveland
Pittsburgh
Harrisburg
Philadelphia
Des Moines
Moline
Fort Wayne
Columbus
Baltimore
Washington
Keokuk
Indianapolis
Springfield
Cincinnati
Kansas City
St. Louis
Charleston
Richmond
Norfolk
Louisville
Wythville
Joplin
Springfield
Raleigh
Knoxville
Nashville
Ashville
Fort Smith
Wilmington
Chattanooga
Memphis
Little Rock
Columbia
Atlanta
Charleston
Birmingham
Jackson
Montgomery
Savannah
Shreveport
Mobile
Jacksonville
New Orleans
Tampa
GULF OF MEXICO
Miami
ATLANTIC OCEAN
−40°
−35°
−30°
−25°
−20°
−15°
−10°
−5°
0°
5°
10°
15°
20°
25°
30°
35°
40°
45°
50°
95°
90°
85°
80°
75°
65°
55°
50°
45°
40°
35°
30°
25°

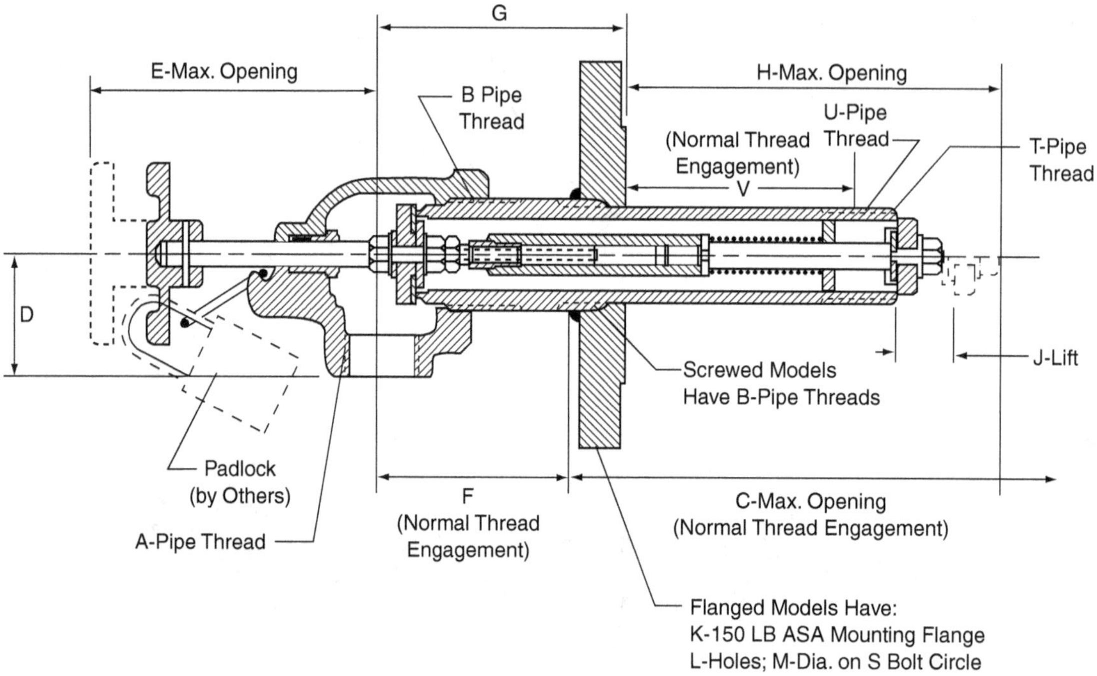

Size	A	B	C	D	E	F	G	H	J	K	L	M	S	T	U	V
2 in.	1½	2	9	2 3/16	5⅞	3⅝	4⅜	8¼	¾	3	4	¾	6	1½	1½	5 3/16
3 in.	2½	3	8⅞	2⅞	5¾	4½	5½	7⅞	⅞	4	8	¾	7½	2	½	4 9/16
4 in.	3	4	8⅞	3⅝	6⅜	4⅞	6	7¾	1⅛	6	8	⅞	9½	3	3½	4¼

Figure 10-3 Double-seating, internal-closing drain valve

It is acceptable to equip a tank with inside ladders and overflow pipes if the tank is known to have a high turnover of warm water. A ladder and overflow can also be installed at the center of the tank and supported by the access tube, as in single-pedestal tanks and extremely large column-type tanks.

The use of interior girders, roof bracing, painter's rails, or virtually any other protrusion below the high-water line or within an area affected by floating or suspended ice is a poor design practice for areas with an LODMT of –20°F (–29°C) or colder. Certain local conditions or tank usage patterns may cause equally severe icing problems in warmer areas.

External Features

Several design issues for the tank exterior are significant in cold-weather operation.

Roof-opening location. Risers or inlet pipes should be directly below roof vents or manholes, or an auxiliary opening should be provided. This arrangement will facilitate thawing the tank if required. No pipe opening should have a protective discharge cap that would preclude the dropping of a probe into the inlet pipe to thaw the ice blockage. Any gratings over piping should not be so restrictive as to prevent thawing lines or pipes from being lowered into the pipe. Gratings also conduct heat and promote freezing.

Static water projections. Unless it is heated and insulated, piping that extends from the tank or from other piping should not contain static water. Drain valves extending on nipples will freeze easily. Drain valves should be of the double-seating, internal-closing type (Figure 10-3).

Additional outlets. Side outlets on the riser pipe for use in pumping in steam or warm water to thaw the riser may be included in the design. These outlets should be plugged at the pipe outer diameter to eliminate an unheated projection.

Clog-resistant vents. Vents should be designed to avoid freezing over or to provide for pressure or vacuum relief. Some tank manufacturers have proprietary designs for this purpose. This type of vent is discussed in chapter 2 and shown in Figure 2-11.

Dark tank color. If a dark color for the tank exterior is aesthetically agreeable and will not keep the water too warm in the summer, the owner should consider using one to absorb radiant heat from the sun.

Riser Pipes

Riser pipes should be properly sized or insulated. For instance, steel water-storage tanks in an area where the LODMT is –20°F (–29°C) usually have small-diameter (less than 12 in. [0.3 m]) risers insulated with at least 4 in. (0.1 m) of polyurethane foam insulation. Large steel-plate risers (36 in. [0.9 m] or larger) have traditionally relied on ice for insulation, but this is no longer recommended. Large-diameter risers (48 in. [1.2 m] and 60 in. [1.5 m]) have been used successfully in areas with the LODMT as low as –20°F (–29°C).

COLD-WEATHER OPERATING PROCEDURES

Even a properly designed tank may freeze under extreme conditions if it is not operated carefully. Water turnover, altitude valves, ground cover, and overflows must all be considered.

Maintaining Water Turnover

With proper pumping management and piping, the heat needed to prevent freezing can theoretically be added to the tank by fresh well water. If 222,000 gal (0.84 ML) of 45°F (7°C) water were pumped in 1 day through the 250,000-gal (0.95-ML) tank mentioned earlier in this chapter (see the Heat-Loss Data section), the heat added would replace the heat lost to the 12-mph (19-km/h) wind in –10°F (–23°C) weather.

Although it is possible to replace most of the water in a tank with warmer water, it is often not the way the tank is operated. Water-storage tanks, particularly elevated tanks and standpipes, because they are connected to the main only by the riser pipe, usually float on the system. As a result, it is possible that a tank can serve a system that uses several times more water per day than the tank capacity yet still receive only a small percentage of fresh water daily. The operating procedures discussed in the following paragraphs will allow a tank to make optimal use of the heat available in incoming water.

Forcing circulation through the tank. If the tank is located near a pumping facility, the tank can be piped so that the new warmer water will be pumped into the tank through one pipe while water is withdrawn from a separate pipe. Ideally, the new water should enter the tank above the capacity midpoint to ensure no short-circuiting of water within the tank. However, this piping arrangement violates a rule of good design practice: any piping extending inside the tank container may be broken off by ice cakes if ice does form. This practice not only minimizes the potential for freezing problems but also ensures better water quality by reducing stagnation.

Pumping management. In a small community or in sections of a large water system, effective water turnover can usually be achieved by pumping management. If the system is automated, a timer can be integrated with the pump controller. In that situation, the controller is set to turn pumps on during the day only when needed to maintain the minimum emergency reserve level. The tank is then pumped to the high level at night. This procedure ensures flow during the normal periods of low demand, and it adds heat at night, when the heat loss is greatest.

Lowering the high water level. If a lower-than-normal water level will still maintain sufficient water for power-outage reserve and fire flow, then during winter the high water level may be lowered about 10 ft (3 m), into the vertical portion of a curved-side tank. This minimizes the probability of ice forming on the upper curved portion or roof support structure. It also minimizes the probability of inadvertently overflowing the tank, thus eliminating the problems cited earlier when tanks overflow.

Timing and sequencing the pumps. In large systems, pumps are usually sequenced cumulatively to meet the demand (that is, only one pump turns on or off at a given time). In winter, system pressure can be allowed to drop a little more; several pumps can then be switched on at nearly the same time. This will cause the water levels in the tanks to fluctuate more than with normal control.

Altitude Valves

In many water distribution systems, altitude valves are used to control the water level in tanks for which the high-water level is at a lower elevation than the pressure gradient of the system. Even some small one-tank systems have been designed with an altitude valve on the tank inlet-outlet line. Both of these are examples of improper use of altitude valves. Altitude-control systems can be designed and installed with timers that force the altitude valve to open, allowing water to flow into and out of the tank and ensuring more frequent turnover.

Altitude valves may malfunction even in good weather. Freezing weather increases the likelihood of malfunction, with frozen pressure-sensing lines giving the altitude valve false signals. This usually causes the tank to overflow, but it may also cause the valve to remain closed, keeping the water in the tank static. Providing electrical heat tape and insulation on the control piping or heating the altitude valve enclosure will minimize these problems.

Ground Cover on the Connection Piping

A common cause of tank freezing is inadequate cover over the pipe leading to the tank. Sometimes soil conditions preclude installing the tank foundation deep enough to provide adequate frost cover; in such cases, fill should be brought in for the cover, or other means should be devised to insulate the pipe.

Site grade. Some tower-type elevated tanks have the riser foundation built higher than the column foundations. The site should be graded higher in the center to prevent the inlet-outlet piping from being exposed to the atmosphere between the concrete riser foundation tunnel roof and the earth.

Compacting fill. If the earth over the connecting piping is not compacted properly, it will settle during the first few years of operation. The ensuing lack of adequate cover, combined with moisture saturation of the depression, creates a potential trouble area for freezing.

Supervising the covering. The base of the tank is usually the location of the interface between constructors, and may become a "no man's land." Unless properly supervised and inspected, the piping may not even get backfilled before the first winter of operation.

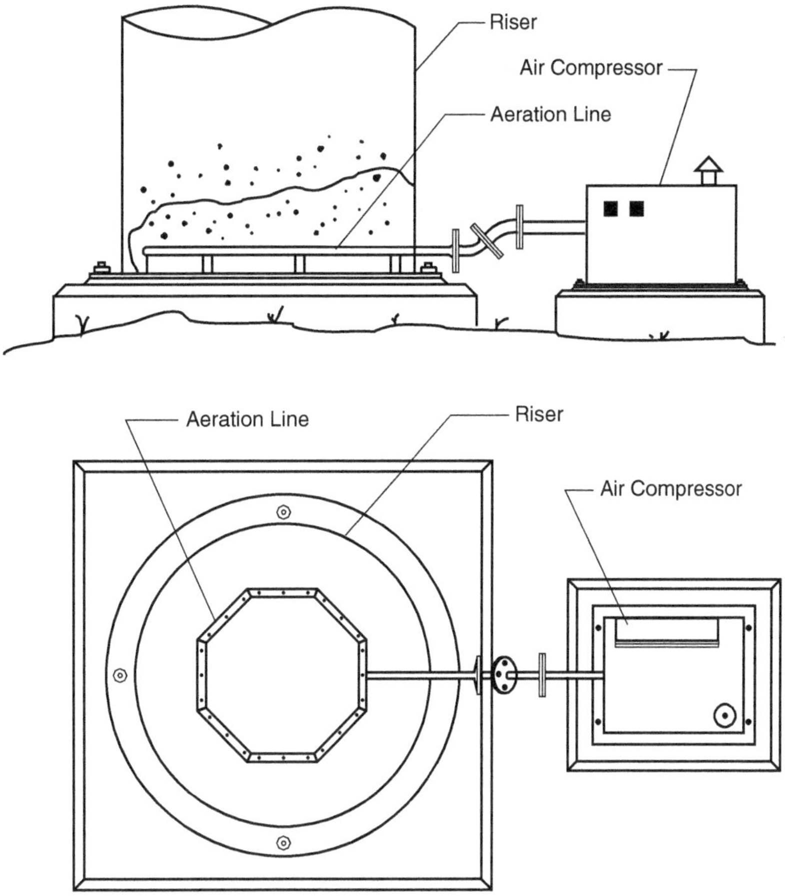

Figure 10-4 Tank riser bubbler system

SYSTEMS TO PREVENT FREEZING

Different types of systems or equipment can be used to prevent tank freezing.

Heating

Heating a community water supply tank is usually not economically feasible, though industrial sprinkler tanks for fire protection have been heated for many years. However, new insurance rate structures and better community water supply systems have allowed many factories either to dismantle the fire protection tank or to discontinue heating it. In many cases, the insurance savings no longer offset the heating energy costs.

Air Bubblers

Air bubbler systems have been used successfully in ground storage tanks and in elevated tanks with large risers. A bubbler system is shown in Figure 10-4. Research on the use of these systems indicates that the compressor used should be one of high volume, with just enough pressure to overcome air-line friction, orifice friction, and the hydrostatic head due to the water depth. Also, an influx of warmer water is required, because the bubbling action tends to remove all the heat from a confined volume of water. The compressor should be equipped with an air filter and water and oil traps to minimize pumping contaminants into the tank.

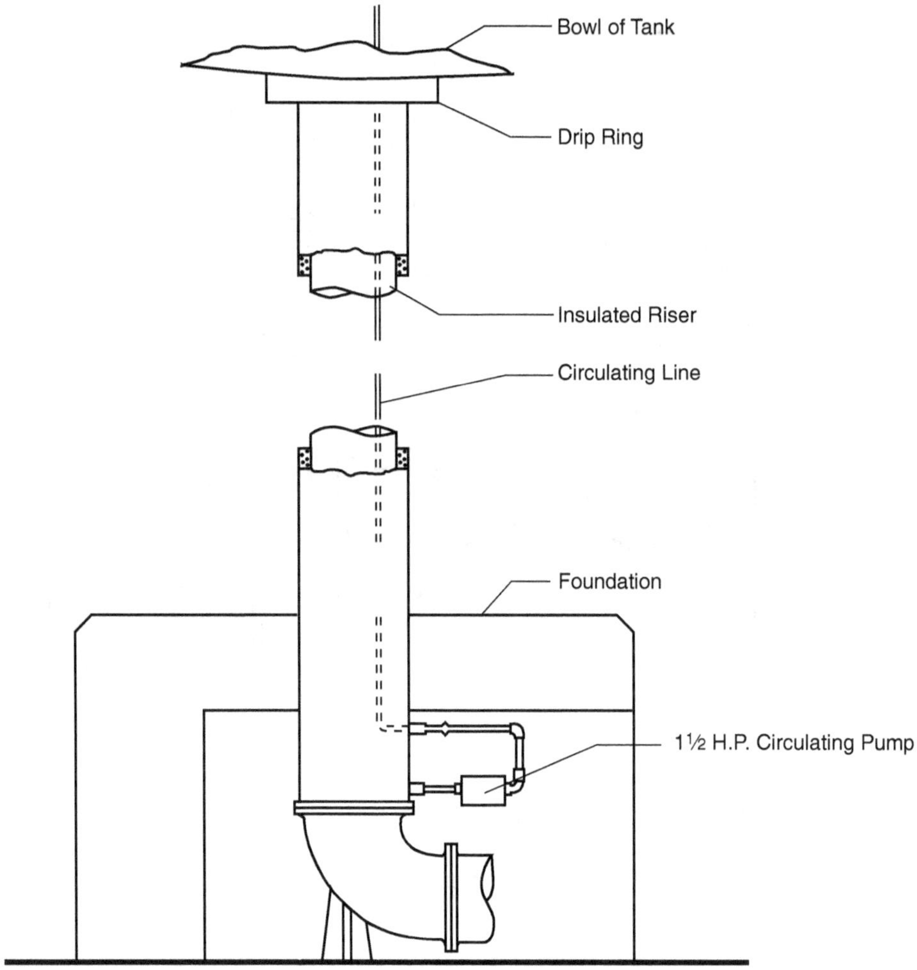

Figure 10-5 Pumped circulation system for small riser pipes

Circulating Pumps

Circulating pumps that do not heat the water have been successful on tanks with small- diameter (6–12 in. [150–300 mm]) riser pipes in Iowa, Minnesota, North Dakota, and South Dakota. A circulating system is shown in Figure 10-5. A relatively small (1.5 hp [1.1 kW]) pump draws water from the base elbow, pulling water down the insulated riser or from the connecting pipe. The pump discharges water into a 1-in. (25-mm) diameter line entering the riser at the base of the tank and discharging into the tank container. This creates a circulation in the riser.

Insulation

As discussed earlier in the context of tank design, it is a common practice to insulate riser pipes. Many ground storage tanks and a few elevated tanks have also been sprayed with urethane foam to insulate them. This yields a rough appearance, however, and problems may arise in maintaining adhesion between the foam and the steel. No matter how thick the insulation, if there is no heat input to the stored water, that water will eventually freeze.

DEALING WITH FROZEN TANKS

Despite all efforts, a tank may freeze. When this happens, there are two ways to thaw the tank: thaw the tank artificially, or wait for warm weather. If in doubt, it is good practice to get advice from a tank company or structural consultant proficient in tank design concerning the dangers presented by a frozen tank.

Piping Damage

If ruptures or leaks are apparent in the inlet or riser piping, it usually is best to wait for warmer weather, or even take additional measures to keep the damaged piping frozen. This will keep water in the tank while things are still frozen overhead.

Damage From Draining

If it is suspected that a tank is frozen, no attempt should be made to drain the tank. Draining could cause damage in three ways:

1. The ice may fall inside the tank, damaging the tank or piping.
2. The vent may be plugged or frozen over, and the roof may collapse (implode) from the vacuum formed by the draining water.
3. The ice may thaw in the tank on only one side (the one facing the sun) This uneven thawing will allow the water to drain off one side of the tank only, producing an eccentric load for which the structure was not designed. In the case of a single pedestal tank, this can cause the support cylinder to fail by bending.

Falling-Ice Hazards

Once ice or icicles form on the tank exterior, they create a danger for people and objects below. If it appears that the ice will cause a structural failure, a distance greater than the height of the tank should be roped off. If the danger appears to be only due to falling ice, an area at least 20 ft (6 m) larger in radius than the tank base should be roped off. If the stability of the structure is in doubt, all precautions should be taken and a structural expert should be called to analyze the situation.

Thawing

Thawing piping. If piping is frozen where it enters the tank or riser pipe at ground level, and it is acceptable to let the water flow again, a tarpaulin or polyethylene film tent can be erected around the base and a gas- or oil-fired portable heater used to heat the area. Similarly, if the tank is of single-pedestal type or other design with a large dry riser, heat may be applied in that portion of the tank.

Thawing tanks. The municipality may attempt to thaw the tank with its own equipment, but contractors are often engaged for the job. A common technique is to lift a ¾-in. (19-mm) diameter heavy-duty hose from the ground up over the top of the tank. The end of the hose terminates with a ½-in. (12-mm) diameter pipe about 10 ft (3 m) long. This pipe is dropped through the tank vent or manhole that is located directly over the inlet or riser pipe. It must be kept slightly off the ice to keep it from sticking to it. Warm water from a fire department tank truck is pumped through the hose, parting the ice as the probe drops down (refer to Figure 10-6). The water must be used judiciously. Stopping in the middle of the process could allow the thawing pipe to freeze in the tank and may require leaving the equipment in the tank for the remainder of the winter.

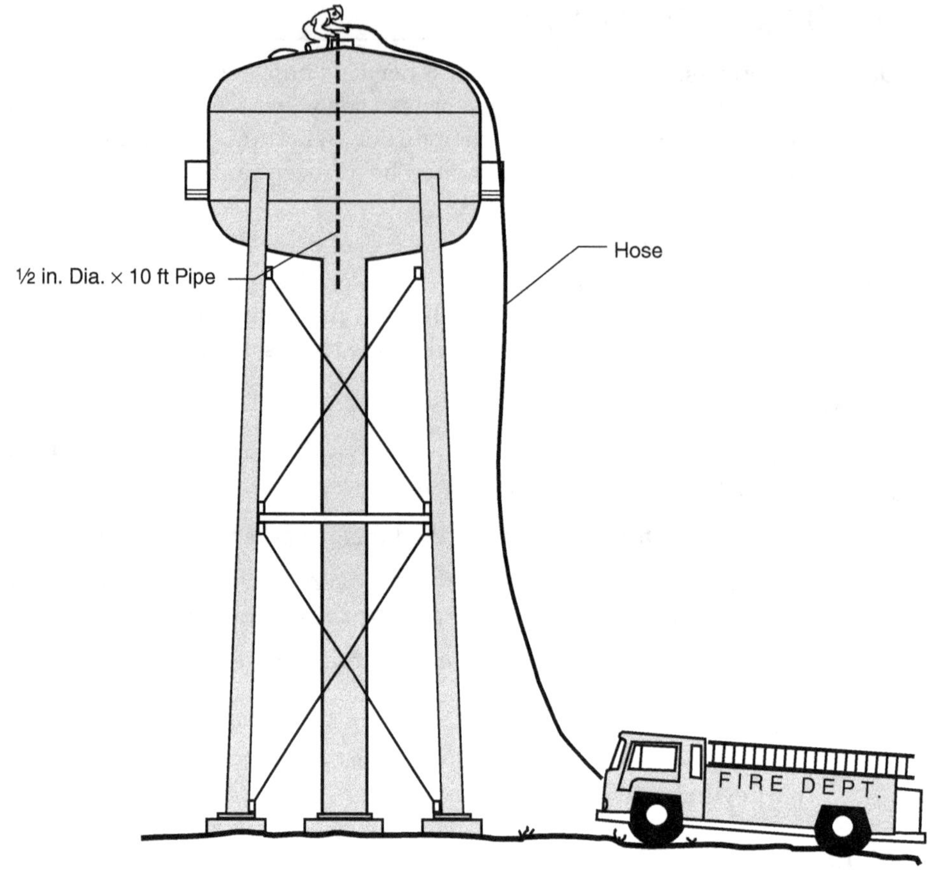

Figure 10-6 Tank-thawing operation

When the tank truck is empty, it must be refilled from utility mains. However, the refill water will not be as warm as the original water in the truck, which was taken from the heated fire station. Steam generators have been attached to this type of probe, but the warm-water method seems to work best, is cheaper, and creates fewer safety hazards.

The same equipment can be used to thaw a frozen riser. However, it may be difficult to thread the probe into the riser pipe. The tank drawings or a recent tank inspection report should be reviewed before this type of riser thawing is attempted, unless the positioning of the piping arrangement is known from experience.

Artificially thawing tanks is expensive and dangerous. In addition, a warm front may move through the day after the crew has thawed the tank, which makes a high thawing bill hard to justify to the utility management.

Repairing Freeze Damage

After the cold weather is over, the tank should be drained and thoroughly inspected for any damage caused by overflowing or freezing. Any modifications recommended to prevent the problem from recurring should be performed while the weather remains warm.

Ellipsoidal and spherical roofs that have been sucked in by vacuum or depressed by a snow load can sometimes be repaired simply. Carefully introducing large volumes of low-pressure air into the tank can sometimes push the roof nearly up to its original contour. This operation should be designed and supervised by an engineer familiar with the structural design of tanks.

Appendix A

Bibliography

ANSI/AWWA C652, Standard for Disinfection of Water-Storage Facilities. Denver, Colo.: American Water Works Association.

ANSI/AWWA D100, Standard for Welded Carbon Steel Tanks for Water Storage. Denver, Colo.: American Water Works Association.

ANSI/AWWA D102, Standard for Coating Steel Water-Storage Tanks. Denver, Colo.: American Water Works Association.

ANSI/AWWA D103, Standard for Factory-Coated Bolted Carbon Steel Tanks for Water Storage. Denver, Colo.: American Water Works Association.

ANSI/AWWA D104, Standard for Automatically Controlled, Impressed-Current Cathodic Protection for the Interior Submerged Surfaces of Steel Water Storage Tanks. Denver, Colo.: American Water Works Association.

ANSI/AWWA D106, Standard for Sacrificial Anode Cathodic Protection Systems for the Interior Submerged Surfaces of Steel Water Storage Tanks. Denver, Colo.: American Water Works Association.

ANSI/AWWA D107, Standard for Composite Elevated Tanks for Water Storage. Denver, Colo.: American Water Works Association.

ANSI/AWWA D108, Standard for Aluminum Dome Roofs for Water Storage Facilities. Denver, Colo.: American Water Works Association.

Beebe, D.G. 1985. Rigging and Ventilation in Tank Painting. *Jour. Prot. Ctgs. & Lngs.*, 2(3):18.

Bushman, J.B. 1983. Cathodic Protection of Water Storage Tanks. In *Steel Water Storage Tanks: What Engineers and Utilities Need to Know.* Sem. Proc. AWWA Annual Conference. Denver, Colo.: American Water Works Association.

Buzck, J.R. 1983. Design of Welded Steel Tanks for Water Storage. In *Steel Water Storage Tanks: What Engineers and Utilities Need to Know.* Sem. Proc. AWWA Annual Conference. Denver, Colo.: American Water Works Association.

Campbell, D.E. 1983. Tank Paint Systems—Selection, Application and Inspection. In *Steel Water Storage Tanks: What Engineers and Utilities Need to Know.* Sem. Proc. AWWA Annual Conference. Denver, Colo.: American Water Works Association.

Christiansen, K.L. 1983. Factory Coated Bolted Steel Tanks. In *Steel Water Storage Tanks: What Engineers and Utilities Need to Know.* Sem. Proc. AWWA Annual Conference. Denver, Colo.: American Water Works Association.

Cull, D.G. 1983. Specifying AWWA Steel Water Storage Tanks. In *Steel Water Storage Tanks: What Engineers and Utilities Need to Know.* Sem. Proc. AWWA Annual Conference. Denver, Colo.: American Water Works Association.

Ellis, C., and J.A. Williams. 1983. Selecting and Sizing Tanks. In *Steel Water Storage Tanks: What Engineers and Utilities Need to Know.* Sem. Proc. AWWA Annual Conference. Denver, Colo.: American Water Works Association.

Ellrich, J. 1983. The Use of Glass Coated Steel Tanks for the Containment of Potable Water. Paper presented at AquaTech of Americas, March 14, at Hershey, Pa.

Knoy, E.C. 1982. Inspection, Evaluation, and Design of Tanks to Minimize Cold Weather Problems. In *Proc. AWWA DSS.* Denver, Colo.: American Water Works Association.

Knoy, E.C. 1982. Maintenance Engineering for Steel Water Storage Tanks. In *Proc. SSPC Symposium.* Pittsburgh, Pa.: Steel Structures Painting Council.

Knoy, E.C. 1983. Inspecting New Tank Construction. In *Steel Water Storage Tanks: What Engineers and Utilities Need to Know.* Sem. Proc. AWWA Annual Conference. Denver, Colo.: American Water Works Association.

Knoy, E.C. 1983. What You Should Know About Tank Maintenance. In *Proc. AWWA DSS.* Denver, Colo.: American Water Works Association.

Knoy, E.C. 1986. Wax-Grease Coatings in Potable Water Storage Tanks. In *Proc. AWWA DSS.* Denver, Colo.: American Water Works Association.

Meier, Stephen W. 2010. *Steel Water Storage Tanks: Design, Construction, Maintenance, and Repair.* Denver, Colo.: American Water Works Association and McGraw Hill.

Roetter, S.P. 1986. Rehabilitation and Maintenance of Steel Water Storage Tanks. In *Water Supply System Rehabilitation.* New York: American Society of Civil Engineers.

Steel Plate Engineering Data. Vol. 1, Steel Tanks for Liquid Storage. Washington, D.C.: Steel Market Development Institute.

Steel Plate Engineering Data. Vol. 2, Useful Information on the Design of Plate Structures. Washington, D.C.: Steel Market Development Institute.

Steel Water Storage. 1984. Cincinnati, Ohio:

Steel Plate Fabricators Association. 1993. *Good Painting Practice. Vol. 1, Steel Structures Painting Manual.* Pittsburgh, Pa.: Steel Structures Painting Council.

Systems and Specifications. Vol. 2, Steel Structures Painting Manuals. Pittsburgh, Pa.: Steel Structures Painting Council.

Visual Standard for Abrasive Blast Cleaned Steel. 1989. SSPC-Vis 1-89. Pittsburgh, Pa.: Society for Protective Coatings.

Appendix **B**

Steel Water Tank Industry Standards Organizations and Information Sources

The standards organizations and information sources listed in this appendix influence the design, fabrication, erection, painting, and maintenance of steel water-storage tanks and their foundations.

STANDARDS ORGANIZATIONS

ACI
American Concrete Institute
P.O. Box 9094
Farmington Hills, MI 48333-9094
www.concrete.org

AISC
American Institute of Steel Construction
One East Wacker Drive, Suite 700
Chicago, IL 60601-1802
www.aisc.org

ANSI
American National Standards Institute
25 West 43rd Street, Fourth Floor
New York, NY 10036
www.ansi.org

API
American Petroleum Institute
1220 L Street, NW
Washington, DC 20005-4070
www.api.org

ASCE
American Society of Civil Engineers
1801 Alexander Bell Drive
Reston, VA 20191
www.asce.org

ASME
American Society of Mechanical Engineers
Two Park Avenue
New York, NY 10016-5990
www.asme.org

ASNT
American Society for Nondestructive Testing
P.O. Box 28518
1711 Arlingate Lane
Columbus, OH 43228-0518
www.asnt.org

ASTM International
100 Barr Harbor Drive
P.O. Box C700
West Conshohocken, PA 19428-2959
www.astm.org

AWS
American Welding Society
8669 Doral Boulevard, Suite 130
Doral, FL 33166
www.aws.org

AWWA
American Water Works Association
6666 West Quincy Avenue
Denver, CO 80235
www.awwa.org

CSA
Canadian Standards Association
178 Rexdale Boulevard
Rexdale, Ontario
Canada M9W 1R3
www.csa.ca

NACE International
1440 South Creek Drive
Houston, TX 77084-4906
www.nace.org

NFPA
National Fire Protection Association
One Batterymarch Park
Quincy, MA 02169-7471
www.nfpa.org

NIST
National Institute of Standards and Technology
100 Bureau Drive, Stop 1070
Gaithersburg, MD 20899-1070
www.nist.gov

NSF International
789 N. Dixboro Road
Ann Arbor, MI 48113-0140
www.nsf.org

SSPC
The Society of Protective Coatings
40 24th Street, 6th Floor
Pittsburgh, PA 15222
www.sspc.org

UL
Underwriters Laboratories Inc.
100 Technology Drive
Broomfield, CO 80021
www.ul.com

USFDA
US Food and Drug Administration
10903 New Hampshire Avenue
Silver Spring, MD 20993
www.fda.gov

TRADE ASSOCIATIONS AND PROFESSIONAL SOCIETIES

AAA
American Arbitration Association
1776 I St., NW, #850
Washington, DC 20006
www.adr.org

AISI
American Iron and Steel Institute
25 Massachusetts Avenue, NW, Suite 800
Washington, DC 20001
www.steel.org

ASCE
American Society of Civil Engineers
1801 Alexander Bell Drive
Reston, VA 20191
www.asce.org

ASNT
American Society for Nondestructive Testing
P.O. Box 28518
1711 Arlingate Lane
Columbus, OH 43228-0518
www.asnt.org

CRSI
Concrete Reinforcing Steel Institute
933 North Plum Grove Road
Schaumburg, IL 60173-4758
www.crsi.org

FM Global
270 Central Avenue
P.O. Box 7500
Johnston, RI 02919
www.fmglobal.com

National Board of Boiler and Pressure Vessel Inspectors
1055 Crupper Avenue
Columbus, OH 43229-1183
www.nationalboard.org

PCA
Portland Cement Association
5420 Old Orchard Road
Skokie, IL 60077
www.cement.org

STI/SPFA
Steel Tank Institute/Steel Plate Fabricators Association
944 Donata Court
Lake Zurich, IL 60047
www.steeltank.com

WRC
Welding Research Council
P.O. Box 1942
New York, NY 10156
www.forengineers.org/wrc

CODES AND STANDARDS

ACI 318, *Building Code Requirements for Structural Concrete*

ASCE 7, *Minimum Design Loads for Buildings and Other Structures*

ASME, *Boiler and Pressure Vessel Code*

AWS D1.1, *Structural Welding Code—Steel*

International Code Council (ICC), *International Building Code*

NFPA 22, Standard for Water Tanks for Private Fire Protection

NFPA 70, *National Electrical Code*

Recommended Standards for Water Works [10 States Standards]
Great Lakes–Upper Mississippi River Board of State Public Health and Environmental Managers

State, provincial, and local building codes should also be consulted.

OTHER INFORMATION SOURCES

Steel producers, tank paint manufacturers, steel tank constructors, and water tank coating consultants are additional sources of information.

This page intentionally blank.

Appendix C

ANSI/AWWA D101-53 (R86)

EDITORIAL NOTE

This appendix consists of a reprint of ANSI/AWWA D101-53 (R86), Standard for Inspecting and Repairing Steel Water Tanks, Standpipes, Reservoirs, and Elevated Tanks for Water Storage. AWWA Manual M42, *Steel Water-Storage Tanks,* replaced D101 as a guide for professionals in the water industry regarding the proper inspection and use of steel water-storage tanks. Although much of the content in D101 still applies today, some is outdated. While reviewing ANSI/AWWA D101-53, readers should keep the following in mind so as not to misinterpret the standard:

Introduction, item 2 (*p. 132*). This step is not always recommended, because the repairs required cannot always be determined at the time of inspection; some additional analysis is usually required. The constructor should visit the tank during a pre-bid or preconstruction meeting.

Introduction, items 5–8 (*p. 132–133*). Separating the repair and recoating work may be a good idea when the repairs are very extensive compared with the coating work required, but it is usually not recommended. Less tank downtime, easier scheduling, and fewer liability problems (such as that arising in holding up one contractor while another is behind schedule) are some of the reasons for combining the repainting and repair contracts. In addition, when the repairs and recoating are combined, the guarantee for the completed job lies in one constructor's hands. Two exceptions to combining the contracts are site work and electrical work. Often these items can be performed by local constructors or water utility personnel at a lower cost than tank repair and recoating constructors can perform them.

Section A-1.4 (*p. 135*). Some inspection firms provide tank-cleaning services as a part of their inspection, so cleaning by the owner is not always essential.

Section A-2.2, first paragraph (*p. 135*). Steel plates can also be inspected with ultrasonic thickness-measuring devices to determine plate thickness. Destructive methods, such as drilling holes, are not required.

Section B-1.3, first paragraph (*p. 139*). It may be advisable to remove spider rods as well as replace them. Most tanks that have spider rods do not require them from a structural standpoint, and the spider rods are often not safe for rigging. However, before the rods are removed, a structural analysis should be performed on the roof to shell juncture to determine if the rods are structurally required.

ANSI/AWWA D101-53 (R86)

AWWA Standard for

Inspecting and Repairing Steel Water Tanks, Standpipes, Reservoirs, and Elevated Tanks for Water Storage

Introduction

This document has been prepared by a joint committee of the American Water Works Assn., the New England Water Works Assn. and the American Welding Society. In developing this document, covering the inspection and repair of steel water storage tanks, standpipes, and reservoirs, the associations make available to their members an outline of methods which, in the opinion of their committees, may properly be followed.

The principles which the committees feel that the tank* owner should follow in repairing a water storage tank are listed below. When the term "owner" is used, it may be taken to mean the person or corporation owning the works; the manager or superintendent of the water department or company; the water board or commission; the city council; or the mayor; that is, whoever is responsible for contracts made by and on behalf of the water utility, whether publicly or privately owned.

1. The tank ower should employ a disinterested inspector to look over the metal surface inside and out, to determine the condition of the paint and interior metal surfaces, and what repairs, if any, are needed. Tanks need repainting frequently, but, if properly maintained, repairs are rarely necessary.
2. The owner should invite any tank repair firm which he wishes to bid on a job to inspect the tank at the same time the inspector does, so that the prospective contractors have full information concerning the conditions.
3. The owner should have the tank drained, cleaned and dry, ready for inspection, on the announced date (See Part A, Sec. A-1.4).
4. The owner should draw up his own proposals for the work to be done and request bids on the work. He should have these proposals approved by his attorney.
5. The tank owner should not combine proposals for repair work and painting in one document. He should always leave himself free to award separate con tracts if he desires.
6. The owner should not permit any bidder to condition one part of the operations upon another part. For example, the owner should not permit a painting contract to bear any conditions, reservations, or cancellations of customary guarantees dependent upon the same bidder receiving the contract for repairs.

* The word "tank" is used hereinafter broadly in place of the lengthy phrase "steel tanks, standpipes, reservoirs, or elevated tanks, for water storage."

7. The owner should, however, include in his proposal a provision that, if the contracts for repairs and for painting are granted to the same contractor, the contractor shall record the amount which he will deduct from the sum of the two independent bids. (The price for the combination job should be less by the stated amount than the sum of the same contractor's bids for separate jobs.)
8. The owner should compare all bids received to see whether a separate bid by one contractor for repair and a separate bid by another contractor for paint ing total less than the adjusted bid in 7 above. If the saving is material and if both the contractors are known to be reliable, separate contracts should be let.
9. The owner, particularly if a municipal department, should arrange for public review of bids and for public letting of contracts.
10. While the use of the contractor's contract forms is not recommended, if the owner chooses to engage in letting contracts for repair or contract forms other than his own, he should first submit them to his attorney, asking him to study them carefully to see:
    a. that no unlimited unit price commitments are included in the contract form;
    b. that no commitments or guarantees by the contractor are conditioned upon any matters not clearly and specifically set forth in the contract;
    c. that no obligations are incurred by the owner which are not specifically set forth in the contract; and
    d. that a clause covering possible extra work is included in the contract form and is of the same general character and equally as specific as that included in Sec. B-2 of Part B of this document.

Referring to 10a above, it is clearly in the interest of equitable business practice that, if the owner and the contractor agree to permit or authorize the contractor to do some unscheduled repair work incidental to painting, this repair work should not be of such extent as to make the painting cost minor by comparison. Again, upon the assumption that a tank owner may choose to disregard the recommendations for prior inspection outlined above, he can properly, and should, stipulate that the total of unscheduled repairs shall not cost more than a given amount, such as the total of the painting contract.

Referring to 10b and 10c above, it is observed that, if extended beyond one year, guarantees of repair or painting work done involve uncertainties which are not necessarily in the interest of either the tank owner or the repair or painting contractor. Frequent and periodic inspection of a water tank is proper and useful, but it should not result in the owner's commitment to a contract for work to be done at an uncertain future date or at a price not arrived at by proper competition.

By the issuance of this document, it is not intended to imply that water departments or companies are not, on their own initiative, competent to develop equitable contracts. Neither is it implied that any member of the associations involved is requested to use the procedure herein outlined. The associations have no power or intent to stipulate that a precise procedure must be followed in any phase of waterworks operation and maintenance. The associations, likewise, are not in a position to police the entire field of contractual operations of water departments and companies. They must content themselves with providing for their members what is considered the best information available on any subject which is documented. It must remain with the responsible executives in the water department or company to protect the interest of the community which they serve.

PART A INSPECTING STEEL TANKS, STANDPIPES, RESERVOIRS, AND ELEVATED TANKS FOR WATER STORAGE

SECTION A-1: General

Sec. A-1.1 Scope

Every steel water storage tank, standpipe, or reservoir should be carefully inspected prior to repair and/or repainting and at any time when leakage or some other apparent deterioration is observed. In any event, all water tanks should be thoroughly inspected at intervals of not more than five years. The standard for painting steel water storage tanks, AWWA D102, includes as an option a first anniversary inspection to be performed by the tank painting contractor. Also, the Foreword suggests that defective areas be repaired annually. It is the intent of this document to define the requisite qualifications of the inspecting agency, the type of inspection to be made and the data considered essential.

Sec. A-1.2 Inspection Service

Under the terms of this recommended practice document, inspection service shall be provided only by organizations or individuals who are properly qualified to do such work. Those so qualified are:

1. An engineering organization whose principals are registered professional engineers, specializing in inspection service and having at least five years' experience in the inspection of steel structures
2. Independent engineers, licensed in the state in which the structure is located, whose practice has included substantial or major attention to steel construction
3. Inspection or safety agencies of the state in which the structure is located, if such agencies are empowered to render inspection service and, further, if such inspection services involve the employment of personnel experienced in steel construction and maintenance.

In all of the above classes of qualified inspection agencies, the inspector or inspectors assigned to the work in the field shall have been properly trained by the organization so qualified and shall have no interest, other than that of a competent inspector, in the performance of any work under consideration at the time the inspection is made.

Sec. A-1.3 Responsibility

The inspector shall assume the entire responsibility for accident to himself while inspecting the structure. He shall make such observations of ladders, railings, roof rods and other parts of the structure as may be necessary to determine their safety for use by him in inspecting the structure. The inspecting company or inspector shall carry adequate worker's compensation, property-damage and public liability insurance and shall fully protect the owner against claims of any nature arising out of the inspection work.

Sec. A-1.4 Draining of Tank

The owner, following proper notification, shall have the tank emptied for the inspector in order that the inside of the roof, sides and bottom of the tank will be properly exposed for inspection. The inside surfaces shall be thoroughly washed down by the owner to remove slime from wall surfaces and loose deposits and dirt from the tank bottom before the inspector arrives. *This is essential.*

Sec. A-1.5 Work Included

The work included under this inspection shall consist of: (1) a field examination; and (2) a specific report upon the structure, using the information form which is a part of this document (Sec. A-2 and A-3) supplemental wherever necessary to fit peculiar local conditions. The inspection work does not include repairs, *except* that, if *vent screens*, cotter pins or nut pins are found to be missing, they shall be replaced at once, or else a special report shall be made promptly to the owner so that he may have the items replaced.

SECTION A-2: Examination and Report

Sec. A-2.1 Condition of Paint

Under the general heading "Condition of Paint," give a description of the condition of the paint as found, stating:

1. Rough approximation of percent of rust area
2. Special locations of such areas, if segregated
3. Character of such rust areas—that is, whether general or blotchy corrosion, loose paint or none.

Sec. A-2.2 Pitting

Determine and report upon the extent and depth of pitting in the area selected, scaled and cleaned by the inspector. The dimensions and locations of each area scaled and cleaned are to be reported. The extent of pitting found should be described both as to location and character. The record of depth of pitting should be as specific as practicable as to location, area affected; whether blotchy, deep, pinpoint or general corrosion. Depth gauges should be used to obtain specific data. Report as in Table A-1.

If plates are badly pitted, report whether drilling holes *or the use of ultrasonic testing* to determine the plate thickness are recommended.

Table A-1 Sample pitting report

Location (by plates numbered from roof down) Percent of Area Affected (approx.)	Max. Depth of Pitting Found	Plate Type of Pitting	Plate Thickness
1			
2			
3			
etc.			
bottom			

Sec. A-2.3 Type of Repairs

Where pitting has penetrated to a depth indicating the necessity for repairs, the report shall so state specifically, describing the location of such spots and their size. If they can be repaired by patches, the inspector shall specify to the extent practicable the size, location and number of patches. If other types of repairs are indicated, the inspector shall specify clearly the type and extent recommended. (Use the same plate reference numbers as in Table A-1.) The inspector, in his report, shall inform the owner of all repair work which he considers necessary. The report need not be limited to the items specifically outlined herein, but should include all items of any nature which the inspector considers material.

Sec. A-2.3 Type of Repairs

Where pitting has penetrated to a depth indicating the necessity for repairs, the report shall so state specifically, describing the location of such spots and their size. If they can be repaired by patches, the inspector shall specify to the extent practicable the size, location and number of patches. If other types of repairs are indicated, the inspector shall specify clearly the type and extent recommended. (Use the same plate reference numbers as in Table A-1.) The inspector, in his report, shall inform the owner of all repair work which he considers necessary. The report need not be limited to the items specifically outlined herein, but should include all items of any nature which the inspector considers material.

SECTION A-3: Detailed Report of Inspector

Sec. A-3.1 Items to Be Reported

In addition to the descriptive report outlined under Section A-2, the inspector shall report on the details itemized below.

3.1.1 *Anchor Bolts.*

1. Are the anchor bolts rusted so as to reduce their strength materially? If so, caliper and record the smallest section. Advise replacement if considered necessary.
2. Are the anchor bolt nuts tight?

3.1.2 *Column Shoes.*

1. Are the column shoes clean and painted?
2. Has dirt accumulated?
3. Are the column shoes seriously rusted? If so, where and to what depth?

3.1.3 *Tower.*

1. Are the tower posts in line?
2. Is there any indication of settlement in the foundations?
3. Are the tower rods in good adjustment and well turned up?
4. Are the tower rods in good condition? If badly rusted, measure the smallest part and report, indicating rods on which reduced section occurs. (Advise replacement if considered necessary.)

3.1.4 *Cotter Pins.*

1. Examine each pin for the presence of cotter pins. Report the location of any pins not so fitted.
2. Where rod pins with nuts are used, advise if the nuts are on with full thread and the end of the thread is well battered.

3.1.5 *Riser Pipe.*

1. Is the riser pipe straight?
2. Are the riser pipe stay rods in good condition?
3. Is the frost casing in good condition and properly supported?

3.1.6 *Indications of Leakage.*

1. Are there any indications of leakage in the riser pipe?
2. In the expansion joint?
3. In the tank proper? If so, give the location and state the type of repair indicated.

3.1.7 *Ladder.* Is the ladder safe?

3.1.8 *Balcony.*

1. Is the balcony safe?
   - ii. Is the balcony floor in good condition?
   - iii. State the amount of rust accumulated on the balcony floor.

3.1.9 *Bolts.* Are any bolts or rivets omitted or missing in the spliced connections of the tower, struts and balcony?

3.1.10 *Paint.* Report on each of the following items:

1. The condition of the paint and metal of the tower
2. The condition of the paint and metal on the outside of the tank bottom, particularly underneath the balcony and post connections
3. The condition of the paint and metal outside of the tank shell
4. The condition of the paint and metal outside of the roof and under the eaves of the roof
5. The condition of the paint and metal inside of the tank shell, based on the areas carefully examined and other information gained by the inspector. (Each sheet shall be carefully examined, removing enough of the accumulation of scale and rust to enable the inspector to report in detail the exact condition of the metal underneath and the extent of rust and pitting.)
6. The condition of the paint and metal on the inside of the roof
7. The condition of the spider and spider rods
8. The finial connection. (Is it solid and safe for the attachment of the painter's trolley?)
9. The condition of the paint and metal on the inside of the bottom of the tank
10. The condition of the paint and metal inside of the riser pipe, particularly at the bottom of the riser pipe.

3.1.11 *Rivets.* Report, by selecting, scaling and examining suitable areas, the condition of the rivets at lap joints and post connections.

3.1.12 *Metal.* Report, by selecting areas to be examined, the condition of the metal between the rivets at the laps and at the post connections.

3.1.13 *Tank Bottom.* Has the bottom of the tank deteriorated because of its having been covered with mud or scale? If so, what conditions are observed? Will scaling and repainting be satisfactory? Are repairs indicated? If so, describe in detail.

3.1.14 *Prior Painting.*

1. When was the tank reported to have been last painted?
2. What material was used?

Sec. A-3.2 Disposition of Report

Two copies of the inspection report shall be delivered to the owner.

It is understood between the owner and the inspection company that copies of the inspection report may be made available by the owner to painting or tank repair contractors, to define the condition of the tank, if bids for repair or painting are desired.

Sec. A-3.3 Payment

Payment for the above-specified services, including all expenses of the inspector, shall be at the lump sum price agreed upon between the owner and the inspector. Payment shall be made within thirty days after the receipt of the inspection report.

SECTION A-4: Cleanliness

The inspector shall conduct all his work in a clean and sanitary manner. No one shall work in a tank if he has been under a physician's care, or has needed a physician's care, within a seven-day period prior to entering or working in the tank. No person shall be permitted to work in a tank who has an abnormal temperature or gives evidence of illness. The tank owner, or a physician employed by him, shall be the judge of the physical fitness or unfitness of any person to enter or work in a tank. No deviation from this stipulation may be permitted.

The tank owner, after work of any nature is done in a tank, is charged with satisfying himself that the tank interior is clean and sanitary before the tank is returned to service. Although a contractor may be required by his contract to clean all surfaces thoroughly before a tank is restored to service, it is the ultimate responsibility of the tank owner either to give the tank a final field inspection or to require such laboratory tests of the quality of water held (for test purposes) in the tank as will demonstrate the good sanitary condition of the tank interior.

PART B REPAIRING STEEL TANKS, STANDPIPES, RESERVOIRS, AND ELEVATED TANKS FOR WATER STORAGE

Introduction

Every tank repair job should be preceded by a detailed inspection of the structure and a report by a competent inspector. Part A of this document deals with inspection. Only from such an inspection is it possible for the tank owner to determine the character and extent of the repairs needed. Without such a determination, it is impossible to define the repair work to be done so that prospective bidders can estimate the cost of the repairs and submit economical proposal therefor.

The owner, using the inspector's report as a basis, shall define and list the repairs to be made. Bids should be made on a fixed-price basis.

It is recognized that specifications for repair work must necessarily be rather general, but the following is recommended as likely to secure good work from responsible bidders at reasonable and definite cost.

SECTION B-1: General

Sec. B-1.1 Work to Be Done

The specifications and contract shall cover all repair work to be done and all compensation to be paid or received therefor. There shall be no other agreements relating either to the work or to compensation.

Sec. B-1.2 Inspection Recorded

The owner shall advise that he has had the structure inspected (naming the inspector) and that a copy of the inspection report is available. The bidders shall familiarize themselves with the report and the conditions of the structure and, upon request to the owner and under such conditions as the owner may prescribe, may make such further inspection as they consider necessary, prior to submitting proposals.

Sec. B-1.3 Work to Be Itemized

The owner shall itemize repair work by units, such as the number of patch plates to be welded to the structure and their approximate average area; the lineal feet of welding or caulking to make the seams watertight; the spider rods to be replaced; the new roof or parts to be replaced, and so forth.*

It is recommended that, in general, welding on tank plates to replace impaired thickness be limited to filling of pits; and that extensive welding over the surfaces of plates or of riveted seams be carefully considered and used only where replacement will not be practicable. All welding performed in the replacement of plates or broken parts, or for other repairs, shall be performed in accordance with the welding provisions of AWWA D100, "Standard for Welded Steel Elevated Tanks, Standpipes, and Reservoirs for Water Storage."

Sec. B-1.4 Lump Sum Bids

Bids shall be on a lump sum basis, including all the contractor's costs, such as transportation, labor, tools, equipment, insurance, delays or other costs of any nature growing out of the repair work above defined.

SECTION B-2: Extra Work

No payment shall be made for any work other than the lump sum amount bid by the contractor, except upon written order of the owner or his authorized representative, *prior to the beginning of the work for which extra compensation is to be requested.* When extra work is so authorized, it shall be based upon an agreed lump sum or unit price, or compensation therefor shall consist of the actual cost of materials used in the extra work; plus the cost of all labor, together with the prorated cost to the contractor for the time actually spent on the extra work by the job foreman; plus rental of equipment, not on hand at the work site, and which is necessary for the extra work, both the necessity of such equipment and its rental cost being subject to acceptance by the tank owner; plus power, fuel, water, and similar operating costs; plus incidental expenses

* Rivets—For a great many years, it was the custom for tank manufacturers to drive tank rivets with flat heads on the inside. After a number of years, a flat-head rivet, which may have rusted to some extent, appears to have lost the major portion of its head, when, in reality, only a small part has disappeared and the rivet has many years of life remaining. Rivets do their work in shear, and as there is very little tension force on a rivet, the head can rust nearly away before replacement is necessary. As long as a rivet stays tight in its hole, it fulfills its function and does not have to be replaced.

expenses incurred as a direct requirement of the extra work, including payroll taxes, workmen's compensation and other insurance based on payroll costs, and any increase in bond premium cost chargeable to the increase in contract amount as a result of the extra work. To the above shall be added 15 percent to cover administrative costs and contractor's profit, which percentage shall be the only amount to be paid in addition to the above-mentioned costs. No claim shall be made for delays caused by the extra work.

SECTION B-3: Watertightness

All work shall be done by experienced workmen, using equipment best adapted to the work. Upon completion of the repair work, the structures shall be watertight. The standard for welded steel tanks for water storage, AWWA D100, requires testing for leaks by filling the tank with water before it is painted. Repair work shall be neatly done. Payment for repair work shall not be made until and unless the job is watertight in the areas in which the contractor was called upon to make repairs to secure watertightness.

SECTION B-4: Responsibility

The contractor shall carry adequate insurance—property, public and employers' liability—and shall protect the owner against suits of any nature which may arise out of work performed by the contractor.

SECTION B-5: Inspection

The owner shall inspect the repair work as it is being done or employ an inspector to do so. Payment for work done shall be made only after the owner has satisfied himself directly that the work is satisfactory; or after an inspector's report has been filed with the owner certifying that the work has been done properly and in accordance with other terms of the contract.

It is in the mutual interest of the owner and the contractor that such inspection be made promptly.

SECTION B-6: Cleanliness

The contractor and all workmen employed by him shall conduct all operations in a clean and sanitary manner. No nuisance shall be committed in a tank; the workmen shall either use proper waste receptacles or leave the tank whenever necessity arises.

The tank owner, after work of any nature is done in a tank, is charged with satisfying himself that the tank interior is clean and sanitary before the tank is returned to service. Although a contractor may be required by his contract to clean all repairs thoroughly before a tank is restored to service, it is the ultimate responsibility of the tank owner either to give the tank a final field inspection or to require such laboratory tests of the quality of water held (for test purposes) in the tank as will demonstrate the good sanitary condition of the tank interior

Index

NOTE: *f.* indicates a figure; *t.* indicates a table.

AWWA Manuals

M1, *Principles of Water Rates, Fees, and Charges*, Sixth Edition, 2012, #30001
M2, *Instrumentation and Control*, Third Edition, 2001, #30002
M3, *Safety Practices for Water Utilities*, Sixth Edition, 2002, #30003
M4, *Water Fluoridation Principles and Practices*, Fifth Edition, 2004, #30004
M5, *Water Utility Management*, Second Edition, 2004, #30005
M6, *Water Meters—Selection, Installation, Testing, and Maintenance*, Fifth Edition, 2012, #30006
M7, *Problem Organisms in Water: Identification and Treatment*, Third Edition, 2004, #30007
M9, *Concrete Pressure Pipe*, Third Edition, 2008, #30009
M11, *Steel Pipe—A Guide for Design and Installation*, Fourth Edition, 2004, #30011
M12, *Simplified Procedures for Water Examination*, Fifth Edition, 2002, #30012
M14, *Recommended Practice for Backflow Prevention and Cross-Connection Control*, Third Edition, 2004, #30014
M17, *Installation, Field Testing, and Maintenance of Fire Hydrants*, Fourth Edition, 2006, #30017
M19, *Emergency Planning for Water Utilities*, Fourth Edition, 2001, #30019
M20, *Water Chlorination/Chloramination Practices and Principles*, Second Edition, 2006, #30020
M21, *Groundwater*, Third Edition, 2003, #30021
M22, *Sizing Water Service Lines and Meters*, Second Edition, 2004, #30022
M23, *PVC Pipe—Design and Installation*, Second Edition, 2003, #30023
M24, *Dual Water Systems*, Third Edition, 2009, #30024
M25, *Flexible-Membrane Covers and Linings for Potable-Water Reservoirs*, Third Edition, 2000, #30025
M27, *External Corrosion: Introduction to Chemistry and Control*, Second Edition, 2004, #30027
M28, *Rehabilitation of Water Mains*, Second Edition, 2001, #30028
M29, *Fundamentals of Water Utility Capital Financing*, Third Edition, 2008, #30029
M30, *Precoat Filtration*, Second Edition, 1995, #30030
M31, *Distribution System Requirements for Fire Protection*, Fourth Edition, 2008, #30031
M32, *Computer Modeling of Water Distribution Systems*, Third Edition, 2012, #30032
M33, *Flowmeters in Water Supply*, Second Edition, 2006, #30033
M36, *Water Audits and Loss Control Programs*, Third Edition, 2009, #30036
M37, *Operational Control of Coagulation and Filtration Processes*, Third Edition, 2011, #30037
M38, *Electrodialysis and Electrodialysis Reversal*, First Edition, 1995, #30038
M41, *Ductile-Iron Pipe and Fittings*, Third Edition, 2009, #30041
M42, *Steel Water-Storage Tanks*, Revised Edition, 2013, #30042-RE
M44, *Distribution Valves: Selection, Installation, Field Testing, and Maintenance*, Second Edition, 2006, #30044
M45, *Fiberglass Pipe Design*, Second Edition, 2005, #30045
M46, *Reverse Osmosis and Nanofiltration*, Second Edition, 2007, #30046
M47, *Capital Project Delivery*, Second Edition, 2010, #30047
M48, *Waterborne Pathogens*, Second Edition, 2006, #30048
M49, *Butterfly Valves: Torque, Head Loss, and Cavitation Analysis*, Second Edition, 2012, #30049
M50, *Water Resources Planning*, Second Edition, 2007, #30050
M51, *Air-Release, Air/Vacuum, and Combination Air Valves*, First Edition, 2001, #30051
M52, *Water Conservation Programs—A Planning Manual*, First Edition, 2006, #30052
M53, *Microfiltration and Ultrafiltration Membranes for Drinking Water*, First Edition, 2005, #30053
M54, *Developing Rates for Small Systems*, First Edition, 2004, #30054
M55, *PE Pipe—Design and Installation*, First Edition, 2006, #30055
M56, *Nitrification Prevention and Control in Drinking Water*, Second Edition, 2013, #30056
M57, *Algae: Source to Treatment*, First Edition, 2010, #30057
M58, *Internal Corrosion Control in Water Distribution Systems*, First Edition, 2011, #30058
M60, *Drought Preparedness and Response*, First Edition, 2011, #30060
M61, *Desalination of Seawater*, First Edition, 2011, #30061

This page intentionally blank.

www.ingramcontent.com/pod-product-compliance
Lightning Source LLC
LaVergne TN
LVHW080528210926
848481LV00022B/3657
* 9 7 8 1 5 8 3 2 1 9 4 8 5 *